Music and Public Health

Lars Ole Bonde • Töres Theorell
Editors

Music and Public Health

A Nordic Perspective

Springer

Editors
Lars Ole Bonde
Department of communication and psychology
Center for Research in Music and Health
The Norwegian Academy of Music
Aalborg University
Aalborg, Denmark

Töres Theorell
Karolinska Institute
Stress Research Institute
Stockholm University
Stockholm, Sweden

ISBN 978-3-030-09439-3 ISBN 978-3-319-76240-1 (eBook)
https://doi.org/10.1007/978-3-319-76240-1

Printed on acid-free paper

This Springer imprint is published by the registered company Springer International Publishing AG part of Springer Nature.
The registered company address is: Gewerbestrasse 11, 6330 Cham, Switzerland

Contents

About the Editors and Contributors

Editors

Lars Ole Bonde, PhD, FAMi is a professor in music therapy at Aalborg University (DK) and professor II at Center for Research in Music and Health (CREMAH) at the Norwegian Academy of Music (N). He is a certified music therapist and clinical supervisor. In the 1980s and 1990s, he worked as music and concert producer. He served as associate editor of the *Nordic Journal of Music Therapy* until 2017. His current research projects include music and public health, music therapy for people suffering from schizophrenia, and monographs on the Danish composers Bent Lorentzen and John Høybye. He has published numerous books, book chapters, and articles on the theory and practice of music therapy, music psychology, music education, music theatre – and music and health.

Töres Theorell, MD, PhD is a professor emeritus at the Karolinska Institute. He served as a clinician (cardiology, internal medicine, occupational medicine, and social medicine) from 1967 to 1990. He became a professor, from 1995 also director, of psychosocial medicine in 1980 at the National Institute for Psychosocial Medicine in Stockholm. At the same time, he was appointed professor of psychosocial medicine at the Karolinska Institute. Since his retirement in 2006, he has served as a scientific consultant at the Stress Research Institute, Stockholm University. His stress research has been focused on physiological mechanisms, epidemiological observations, and controlled intervention studies. He has been doing research on culture and health since the 1980s, mainly in the music area. He has written more than 450 articles in international scientific journals and authored and edited several books, including *Psychological Health Effects of Musical Experiences: Theories, Studies and Reflections in Music Health Science* (2014).

Contributors

Anne Haugland Balsnes, PhD is a professor of music and manager of research at University of Agder and Ansgar University College at Kristiansand, Norway. Her PhD from the Norwegian Academy of Music was based on the thesis *Learning in Choirs – Belcanto as a Community of Practice* (2009). This study was followed by the postdoctoral project *Choral Singing for a Better Life*, which resulted in the book *Choral Singing – Ideal for Human Fellowship?* (2014). In addition, Balsnes has written several articles on choral singing and health in journals and anthologies. She is also a practicing conductor and singer.

Margarida Baltazar, PhD studied clinical psychology in the University of Lisbon. She has worked as a psychologist and as a piano teacher, and the motivation to conciliate the sciences of the mind and music has been long present. She completed her doctoral studies at the Department of Music, Art, and Culture Studies of the University of Jyväskylä. Her research interests include affect self-regulation through music, musical emotions, and wellbeing. Her work has been focusing both on theoretical and daily life implications of music use.

Kari Bjerke Batt-Rawden, PhD is an associate professor at the Institute of Health Science, Norwegian University of Science and Technology (NTNU). She is head of the master study in "Health Promotion and Community Care" at NTNU. Batt-Rawden also belongs to the Sociology of the Arts Group [SocArts] at Exeter University and the Research Group "Health Promoting Communities" at NTNU in Gjøvik in conjunction with Center for Health Promotion Research at NTNU in Trondheim. She has published several international, scientific research papers on links between music, health, and quality of life, mainly based on the use of qualitative methods and a salutogenic perspective.

Lars Rye Bertelsen, DMTF music therapist is a doctoral student in the Doctoral Program of Music Therapy at Aalborg University, Denmark. He founded and runs a private music therapy clinic community since 1999, and holds a position as a clinician and research assistant at the Music Therapy Research Clinic at the Aalborg University Hospital – Psychiatry since 2005. He is co-inventor of "The Music Star" application, and he regularly gives lectures and workshops on music as communication, as well as being a part-time professional musician.

Ola Ekholm, Msc is a senior advisor at the National Institute of Public Health at the University of Southern Denmark, Denmark. His main research interests are in the areas of using population-based health surveys in epidemiological studies. He has authored/co-authored more than 85 research papers in international peer-reviewed journals. He is also a supervisor for several PhD students.

Tora Söderström Gaden, MA-MT music therapist graduated from the Norwegian Academy of Music in 2015 with the masters thesis *Song, Interplay and Bonding* based on her music therapy practice project at a public health clinic with first-time mothers and their young infants. She is currently working at Akershus University Hospital's Pediatric Department, where she is also leading a project on adapting international models and methods for music therapy in neonatal intensive care in the context of Norwegian health care and neonatal intensive care units. From 2018 PhD student at GAMUT, Bergen.

Eva Bojner Horwitz, PhD, RPT, and Reg DMT is a medical doctor, a cultural health researcher, and specialized in psychosomatic medicine and creative arts. She is a co-founder and director of education at the Center for Social Sustainability, CSS, Karolinska Institutet (KI) Sweden and researcher at the Department of Clinical neuroscience KI and Department of Public Health and Caring Sciences, Uppsala University. She is anchored in interdisciplinary research, has doctoral students, authored scientific articles, books, and book chapters. Her research expertise covers interventions with dance/music/theater, alexithymia, creative achievement, embodiment, emotional regulation, nature-related activities, fairy tales, public health, and quantitative and qualitative research methods.

Stine Lindahl Jacobsen, PhD is an associate professor and head of the music therapy MA Program at Aalborg University, Denmark. She co-founded and runs a cross-sector knowledge and research center for culture and health in Denmark (Nordjysk Center for Kultur og Sundhed). Jacobsen regularly lectures at universities in Germany, Austria, Spain, and Norway. She has published various articles and chapters in the area of families at risk, standardized assessment tools, and effect studies. Jacobsen actively takes part in public and political debates on the use of music and other art disciplines to promote health.

Steinar Krokstad, PhD is a professor of social medicine and general manager of the HUNT Research Center, Department of Social Medicine and Nursing, Faculty of Medicine and Health Sciences at NTNU. He is a former specialist in general medicine and now a specialist in psychiatry and general practitioner at the Clinic for Mental Health and Substance Abuse, Levanger Hospital, Helse Nord-Trøndelag HF. He is the deputy chairman of the National Center for Cultural Health and Care in Norway. He holds a doctorate in social epidemiology/social medicine, and his research projects include socioeconomic causes of bad health, illness, mortality, and disability plus research in psychiatric epidemiology and national health services. He has contributed several book chapters in community medicine, social medicine, and medical sociology. He has been heavily involved in health prevention and promotion, in cooperation with local, regional, and national authorities in Norway.

Bente Irene Løkken is a PhD fellow in culture, activity, and mental health at the faculty of medicine and health science at the Norwegian University of Science and Technology (NTNU), and employed at Nord University, faculty of nursing and health science. Her current fields of interest are public health, epidemiology, and register data. She is currently lecturing on the subject public health and health-related behavior. The title of her PhD project is "Leisure time cultural activities in the population; mortality, sickness absence and use of health care - cohort analyses from the HUNT 3 Survey (2006-08)". The main objective of the PhD is to investigate how attendance in leisure time cultural activities, both receptive and creative, affect different aspects of health in the population: mortality, sickness absence, and use of health services.

Helle Nystrup Lund, DMTF music therapist and PhD student at the Unit for Psychiatric Research at Aalborg University Hospital, Denmark. Her area of research is depression and music listening to improve sleep quality and to promote health. She is an experienced teacher and clinician, and co-inventor of the Music Star app. She publishes in the area of music therapy and music listening and has presented to international audiences for health professionals. She is also a professional jazz pianist and composer, performing with her jazz trio "Helle Lund Trio." Combining workshops and performances she has toured in Norway, Finland, Scotland, and France.

Dafna Merom, PhD is a professor of physical activity and health at Western Sydney University, NSW Australia. She is an internationally known expert in the area of physical activity epidemiology, measurement, surveillance, and promotion. Merom has been awarded more than $3.5million research grants, including being a recipient of the National Health and Medical Research postgraduate award. These grants have been used to develop and evaluate interventions to promote active living to various population groups and in various settings including the first international large-scale study on the effect of social dancing on falls and cognition among older adults.

Vegar Rangul holds a PhD in public health and general practice, behavioral epidemiology. He is a specialist in physical activity epidemiology and has extensive experience with the use of large epidemiological data sets to resolve issues tied into the health effects of physical activity and its interaction with general health behavior and cardiovascular disease risk in particular. He has expertise in behavioral epidemiology, measurement of physical and cultural activities, and epidemiological population studies. He is the chairman of The Norwegian Centre of Arts and Health and project leader of the National Educational Program in music-based environmental treatment, with integrated use of music, song, and movement.

Hanne Mette Ridder, PhD, DMTF music therapy supervisor, is a professor and head of the Doctoral Program in Music Therapy at Aalborg University, Denmark. She has long clinical experience in dementia care, leads research on music therapy in dementia and neurocognitive disorders, and is engaged in various international research networks. She has published extensively, and recently edited a Danish/Norwegian book on music therapy in healthy aging – *Musikkterapi og eldrehelse* – together with Dr. Brynjulf Stige.

Suvi Saarikallio, PhD Docent of Music Psychology, works as a senior researcher at the Center for Interdisciplinary Music Research, Department of Music, Art and Culture Studies, University of Jyväskylä, Finland. She is an expert in music psychological research on emotion regulation, youth psychosocial development, and emotional wellbeing. After completing a PhD in Music Education in 2007, Saarikallio has rapidly established an internationally acknowledged research career combining approaches from music psychology, music therapy, and music education, using methods from qualitative theory development to motion capture, neural measures, and psychometrics, and actively publishing her work in the top journals of the field.

Karette Stensæth, PhD is professor in music therapy at the Norwegian Academy of Music, where she also works as the director of the Centre for Research in Music and Health (CREMAH). Her research interest is broad and involves social and philosophical perspectives on music and health. Stensæth has edited several books in the CREMAH Series. Her fresh monograph is on responsiveness in music therapy improvisation inspired by Mikhail Bakhtin's ideas. Stensæth has much clinical experience as a music therapist working with children and youth with special needs. She sings, plays, and composes music in her spare time.

Brynjulf Stige, PhD is a professor of music therapy at the University of Bergen and Head of GAMUT – The Grieg Academy Music Therapy Research Centre, University of Bergen & Uni Research Health, Norway. Stige has founded two international peer-reviewed journals: *Nordic Journal of Music Therapy* and *Voices: A World Forum for Music Therapy*. He is currently the founding leader of POLYFON knowledge cluster for music therapy, a partnership that enables collaboration between research institutes, hospitals, municipalities, and counties on service development, research, education, and dissemination. Stige's research explores relationships between music therapy, culture, and communities of practice.

Erik R. Sund, PhD is a research scientist at the department of public health and nursing, NTNU, and an associate professor at the faculty of health sciences and nursing, Nord University. He has a PhD from NTNU on the topic social and geographical inequalities in health. He has a wide range of research interests within public health but with a particular interest in statistical modeling. He is currently involved in research on gene-environment interactions and cognitive and mental health in the elderly.

Gro Trondalen, PhD, special education teacher, music therapist, fellow of AMI, is a professor in music therapy and former director of the Centre for Research in Music and Health (CREMAH) at the Norwegian Academy of Music in Oslo, Norway. She is an experienced music therapy clinician and supervisor, and maintains a private practice in The Bonny Method of Guided Imagery and Music (GIM).

Fredrik Ullén, PhD is a professor of cognitive neuroscience at the Department of Neuroscience, Karolinska Institutet, since 2010. His research focuses on the neuropsychology of expertise and creativity, i.e., the various brain mechanisms that allow us to perform at a very high level within a specific field, using music as a model domain. Methodologically, his team combines neuroimaging with experimental psychology and behavior genetic analyses. He is currently heading a larger research program *Humans Making Music* that involves collaborations with the Swedish Twin Registry and other research groups both within and outside Sweden. In addition to his career as a scientist, Ullén is active as a professional pianist. Professor Ullén is a fellow of the Swedish Royal Academy of Music (2007) and Academia Europaea (2017).

Chapter 1
Introduction

Lars Ole Bonde and Töres Theorell

History

In November 2011, Copenhagen hosted the first conference on music and public health held in a Nordic country. The European Public Health Association (EUPHA), Association of Schools of Public Health in the European Region (ASPHER) and the Danish Society of Public Health organized the fourth European Public Health Conference in Copenhagen 10–12 November, with more than 1300 participants and many hundred presentations. The specific event at the Royal Library on 9 November was an invited preconference (1 of 20), arranged in a collaboration between the Department of Communication and Psychology/Music therapy at AAU and the Center for Research in Music and Health (CREMAH) at the Norwegian Academy of Music, Oslo.

'Music as/in therapy' is well established as an evidence-based treatment modality all over the world, so there is a solid knowledge base documenting how and why music can help people with physiological, psychological, existential, spiritual and social problems and pathologies. In the context of the present book, 'music therapy' is defined as music interventions and services delivered by a qualified/certified music therapist. The context is often clinical, with the aim of promoting the health of patients or clients with mental or physical diagnoses. However, 'community music therapy' is often practiced in broader, non-clinical contexts and with undiagnosed clients or citizens. 'Music and health' is a wider field where the use of music experiences to promote health and wellbeing in everyday life is studied and promoted, also by other agents than qualified music therapists, e.g. nurses, social

L. O. Bonde (✉)
Department of Communication and Psychology, Aalborg University, Aalborg, Denmark
e-mail: lobo@hum.aau.dk

T. Theorell
Karolinska Institute, Stress Research Institute, Stockholm University, Stockholm, Sweden
e-mail: tores.theorell@su.se

L. O. Bonde, T. Theorell (eds.), *Music and Public Health*,
https://doi.org/10.1007/978-3-319-76240-1_1

workers or musicians. 'Music and public health' is a new, interdisciplinary field where social scientists, medical doctors, music psychologists, music therapists, musicologists and health professionals are creating a knowledge base for the focused application of music experiences and activities in a public health perspective.

The purpose of the 2011 preconference was to present a state of the art by three international keynote speakers and to give an overview of the Nordic perspectives on music as health promotion and to discuss problems, achievements and ideas.

The programme was divided into three sections: (1) international perspectives on music and public health; (2) perspectives on music and public health as seen from the Center for Research in Music and Health, Oslo; (3) Scandinavian perspectives on music and health, as seen by leading researchers from Norway, Sweden, Finland and Denmark. Different theoretical and practical models were presented, and recent research results from clinical and non-clinical areas were related to the public health perspective.

For decades, keynote speaker Suzanne B. Hanser (Berklee College of Music, USA) has studied how music therapy can assist in the fields of childbirth, depression, and cancer. She examined how evidence-based strategies developed in her clinical practice and documented in research can be translated to the general public. This is exemplified in the book *Manage Your Stress and Pain through Music* (Hanser & Mandel 2010). Hanser sees one important role of music in modern health care in bringing homeostasis to the autonomic nervous system, and her vision is the integrating of music therapy into mind-body approaches, giving it a role in the new science of integrative medicine.

Raymond MacDonald (Glasgow Caledonian University, Scotland) is a professor of music psychology and also a saxophone player specialized in free improvisation. In his keynote he presented an overview of current conceptions of improvisation, highlighting a number of key themes in relation to improvisation and musical identities within a health-care context. Musical identities refer to the multitude of ways in which interactions with music (both listening and playing) can influence our sense of self, and MacDonald demonstrated how participation in improvisation workshops can have health benefits for cancer patients, highlighting the potential of music activities as innovative psychological interventions in a health-care setting.

Stephen Clift (Sidney De Haan Research Centre for Music, Arts and Health, UK) set out from the fact that the WHO Commission on Social Determinants of Health, under the leadership of Michael Marmot, gave no consideration to the role of music, or the wider field of creative arts, as potential contributors to positive health and wellbeing. Clift presented evidence from many studies documenting how group singing can promote psychological and social wellbeing and help society to meet a number of key challenges linked to an increasingly elderly population and the growing burden of long-term conditions. The big challenge is how to organize such activities on a sufficient scale to have relevance for public health and to assess potential cost savings to health services from a health economics standpoint.

Researchers from the Center for Research in Music and Health, Oslo (Even Ruud, Gro Trondalen, Karette Stensæth and Torill Vist) gave examples of how the center works to increase public knowledge and awareness of the health potential of music activities, in everyday life as well as in clinical and community work, with people suffering from health deficits. Researchers from Sweden (Lasse Liliestam), Norway (Brynjulf Stige), Denmark (Hanne Mette Ridder) and Finland (Suvi Sarikallio) reported from research studies documenting the health benefits of singing, playing and listening to music for diverse non-clinical groups, including adolescents and the elderly, and for people with physical or mental problems.

'Health musicking' was highlighted as a concept integrating the many different perspectives and results of the studies presented (Stige 2002; Stige and Aarø 2012; Bonde 2011). To take part in musical activities and share experiences with other people is a resource with a health dimension and potential well documented in small studies from music therapy, music psychology and music ethnology.

The preconference provided a sharing of promising results and ideas which concluded that the big challenge in a public health perspective is to transform the existing knowledge into practical initiatives in the field of health prophylaxis and prevention. This requires large controlled studies, not only correlational but also longitudinal and cohort studies, and thus, cross-disciplinary networking and funding in a much bigger scale than now is necessary.

The 2011 event was followed up in two ways in the following years. In Denmark, a collaboration between Music Therapy AAU and the National Institute of Public Health (SDU) was initiated, leading to the research project documented in Chap. 2 of the present book. In the Nordic countries, different initiatives provided a continuation and expansion of the new field. The most important initiatives were a conference on Culture and Public Health in Stockholm in November 2014, documented in an anthology (Bojner Horwitz et al. 2015), and the formation of a Nordic Network for Research in Music and Public Health at a CREMAH seminar in Copenhagen in August 2016. Many of the researchers participating in these events have contributed to the present volume, and they also presented a round table at the EUPHA conference in Stockholm, November 2017. All in all, this is a very promising story about how researchers with very different backgrounds have found together to develop the truly interdisciplinary study of music and public health.

Part One: Epidemiological Research in Music and Public Health in the Nordic Countries

In some Nordic countries – especially Sweden and Norway – there has been a tradition of investigating the influence of engagement in cultural activities and events on health; however, until recently there has not been a specific focus on music behaviour and music as a health resource. The three chapters in part one present new, empirical researches from Denmark, Norway and Sweden, with focus on association between music and health.

Chapter 2 (Ekholm and Bonde) presents a Danish epidemiological study of music as a health resource in the life of adult Danes. Data originates from the Danish

Health and Morbidity Survey 2013, the latest survey in a long row. The survey was based on a random sample of 25,000 adult Danes (response rate, 57%). Besides standard health-related questions, the survey included for the first time eight specific questions addressing the informants' music making, their use of music in daily life and beliefs of music as a health resource. The results include an overview of musical behaviours and beliefs in the adult population and an examination of associations between singing/playing and various health-related outcomes such as quality of life and mental and physical health. Finally, the study takes a closer look at self-rated health of adult amateur musicians and professional musicians versus nonmusicians.

The study revealed significant differences in health behaviour between nonmusicians, amateur musicians and professional musicians. It was also documented that active amateur musicians reported a significantly better health than all other groups. A clear association was found between attending live concerts and reporting good health, and also believing that music activities and experiences can help to stay healthy.

Chapter 3 (Løkken et al.) reports recent findings from the Norwegian HUNT study, which are ongoing, large-scale investigations of health in the region of Trøndelag. Creative activities, like playing an instrument, singing or creating theatre performances, can affect biological processes in the human body and have shown to have a positive health effect on patients. Previous studies from the Nord-Trøndelag Health Study (HUNT), Norway indicate that people who are culturally active experienced better self-reported health, was more satisfied with their lives and experienced less anxiety and depression.

The chapter focuses on the association between engagement in performing music, singing and acting with self-rated health and all-cause mortality in the population. The main research question is whether people who actively engage in music, singing and acting have better self-rated health (SRH) and survive longer compared to those who do not participate in these activities on a regular basis. Secondly, possible gender differences are explored. Preliminary results suggest that women not engaged in playing an instrument, singing or acting had an increased risk for having poor self-rated health compared to women who actively participated in such activities. Men not engaged in playing an instrument, singing or acting had increased risk of dying compared to men who actively participated. In summary, these findings suggest that participating frequently in music, singing and acting appears to increase subjective self-reported health for women and reduce all-cause mortality for men. Stimulating such activities may have positive health effects in the population.

It is of great importance for social interactions and for wellbeing, and consequently also for health, that a person is able to interpret and differentiate, as well as to describe and communicate emotions. In psychosomatic medicine, difficulties in such abilities are labelled *alexithymia* (introduced by Sifneos, 1973, 1996). An alexithymic person has difficulties in the communication of feelings. This may lead to poor communication with others and also to reduced capacity for fantasizing and symbolic thinking.

Chapter 4 (Theorell and Ullén) takes as a point of departure that questions about possible relationships of repeated musical experiences to emotional skills have not been studied extensively. The chapter explores two studies of such relationships and also the extent and nature of such relationships in relation to health and creativity. The main research question is whether there is any correlation between a life with music on one hand and the ability to handle emotions on the other hand.

The study sample was recruited from the Swedish Twin Registry. The participants were in the ages 27–54. All analyses were adjusted for age. In the genetic analysis, there were 8110 subjects who had valid information about zygosity, alexithymia score and number of hours of music practice. An interesting finding in the first study was that those who reported a high number of hours of music practice and who had practiced ensemble had a particularly low risk of having a low alexithymia score. A high level of musical activities throughout life, particularly when including ensemble playing or singing, was associated with a lowered level of alexithymia. In other words, musicians are in general and particularly if they have had extensive experience of ensemble playing or singing better at handling emotions than others.

The second study examined the statistical correlation between achievement in cultural activities in general and alexithymia. The results showed both for men and women that high achievement in writing and music contributes statistically independently of one another and of other artistic achievements to a low alexithymia score. Together, the studies based upon a large cohort of twins from the Swedish Twin Registry have shown that musicality and ability to handle emotions are interrelated and that these variables are associated with likelihood of working in creative occupations. The patterns are different for men and women.

Part Two: Empirical Studies as a Basis for Theories on Music and Public Health

The public debate about music's possible role in public health has sometimes been confused because discussants in the public debate have diverging definitions of the concepts they are using and also widely different theories underlying their statements (Theorell 2014; Bonde 2015). We, therefore, feel that it is necessary to base our discussions regarding relationships between music experiences and health by introducing theories based upon empirical research.

Chapter 5 (Theorell) presents knowledge on music biology relevant to public health. It draws on results from a number of research projects that illustrate how researchers reflect on links between music and public health. Music enters the brain in a different way than do conversations based on words, and since the brain tends to react more directly and rapidly to music, this may sometimes create a basis for surprise and unexpected reorientation in life. The body seems to react in specific ways to specific emotions – resulting in various combinations of psychophysiological states (dilated or constricted arteries, increased or decreased variation in heart rate, accelerated or decelerated pulse, elevated or lowered blood pressure, increased or decreased sweating, etc.) when music induces or amplifies emotional states. In addition, the body adapts its hormones and its immune system to the musical experiences.

Experiments with school children have shown that musical collective experiences (having fun with music together and making pupils collaborate with one another) can contribute to an improved social environment possibly favouring learning at school. A calmer atmosphere is mirrored in reduced saliva cortisol levels. Modern recording techniques have made it possible to record immediate online physiology during musical experiences, for instance, in the gastrointestinal system (changes in peristalsis), in breathing patterns and in the arteries as well as during intense experiences such as flow and goose skin. Music in the gym, during choir singing and in clinical applications such as choir singing for patients with chronic respiratory disease is discussed in the chapter. One conclusion is that there is extensive knowledge about immediate reactions during music experiences, but that long-term biological consequences of repeated musical experiences (such as choir singing or instrument playing in groups once a week for years) have been understudied although such research is beginning to emerge.

Chapter 6 (Ridder) presents theories on healthy aging and prevention of behavioural and psychological symptoms of dementia. First of all, general concepts like healthy aging, personal growth and loneliness are discussed from general societal points of view. After this, the terms growth and cognitive reserve are discussed. According to Erikson, old age is a period for achieving integrity, which contributes to wisdom and is opposed to stagnation and despair. It is argued that cognitive reserve to a great extent determines the existential consequences of aging, in particular premature aging of nerve cells. It has been observed that individuals who are able to perform well in cognitive tests in spite of severely aged brain cells have a better connected brain, with mechanisms able to reorganize around the Alzheimer disease pathology. This is the main element in cognitive reserve. Biological observations indicate that an important vehicle in cognitive reserve is norepinephrine, which is upregulated by the crucial factors education level, mental activity, social engagement and enriched/novel environments.

It is emphasized that sensory decline is a neglected part of dementia. Disturbed sensory processing makes it very challenging to understand sensory input and to make sense of situations and interactions. This is in particular the case where listening to and practicing music gets into the picture. To perform and to actively listen to music is a complex cognitive process, and it is argued that this may serve as training for the brain with beneficial effects on daily functioning. Practicing a musical instrument involves movements of parts of the body in a temporal context, as does dancing.

Music is described as having the power to unlock memories and other cognitive capacities in Alzheimer's disease. The regions normally involved in musical memory encoding are strikingly well preserved in Alzheimer's disease. The conclusion is that listening to and performing music has a great potential for preserving functions in old age, even among subjects with Alzheimer brains.

Chapter 7 (Saarikallio and Baltazar) introduces and discusses the concept 'social-emotional health' which is fundamental for the healthy adaptation to the environment. Good social-emotional skills are vital throughout the lifespan, and they can predict mental health and adaptation to society. The development of these skills is especially relevant during adolescence. During these years, music listening serves as a forum for peer group identification and for friendship formation and stability. Social-emotional competence is a positive and wider concept than alexithymia (discussed in Chap. 4). Social-emotional competence is an amalgamation of two approaches to noncognitive competence – emotional intelligence and social intelligence. The authors state that despite music's intrinsic connections to social and emotional experiences, the evidence base linking music engagement to general social-emotional competence is still relatively sparse. But several interesting examples from recent research are presented. For instance, in an experimental study with random control design, musical training made 6-year old children better than their peers in the control group to identify difficult emotional expressions in speech prosody.

The strongest link between music and social-emotional competence in research so far has been through emotion regulation. The observed interrelations between the used strategies and the observed symptomatology point to favouring the use of music for self-reflection, reappraisal and distraction in contrast to avoidant and ruminative emotion regulation strategies. The authors suggest that the concept social-emotional competence could be used as a theoretical framework to clarify the conceptual divergence in the field.

Chapter 8 (Stige) on 'health musicking' introduces a wide view of public health. The 'public health' concept underlying this chapter includes but goes beyond population prevention of disease. It is pointed out that public health strategies are based in the human rights, so that social justice, participation and empowerment are integral rather than additional goals.

Extensive experiments with strategies for implementation of music therapy within mental health services in Bergen, Norway, are described. In Norway in 2015, a new initiative came from the government: all health trusts were instructed to develop medication-free services to patients preferring such treatment. The health trust in Bergen with public health responsibility for several hundred thousand people is the first health trust with systematic implementation of music therapy services. In this process collaboration with local musicians, organizations and culture authorities is central.

The concept 'health musicking' is introduced to communicate the idea that relationships between music and health could be understood as situated processes of participation. Such processes evolve inside and outside conventional music therapy practices. Health musicking is defined as the appraisal and appropriation of the health affordances of the arena, agenda, agents, activities and artefacts of a music practice. The necessary components *arena, agenda, agents, activities* and *artefacts* are discussed. The importance of reciprocity in musicking is emphasized.

Participation in a social and situated activity is considered central to our understanding of how music could be relevant for human health.

Despite an increasing amount of primary and review studies in the field, it still remains uncertain how and under what conditions change strategies and interventions most effectively can be translated and exchanged to health professionals and integrated in their organizations. Knowledge regarding how governance is managed and carried out in health-care services and public health work is needed. In such processes, partnerships might represent alternatives to vertical hierarchies and horizontal demarcations, but they are not without their own limitations and problems. Experiences from the so-called POLYFON knowledge cluster for music therapy are described and discussed.

Chapter 9 (Stensæth) on music *as* participation is devoted to questions regarding music's potential to reduce isolation and thereby improve public health. The point of departure is that social isolation and loneliness are about to become the biggest public health threat of our age. There is a great need to explore new ways to fight social isolation and to find meaningful ways for people to be with others while engaging in participation, both on an individual level and on a community level. Music, especially through the development of the discipline of music therapy, tries novel ways of approaching some of the challenges connected to social isolation.

It is emphasized that loneliness and social isolation are not individual phenomena. Social isolation is characterized by an absence of social interactions and social structures and engagement with wider community activities or structures. Loneliness refers to an individual's personal, subjective sense of lacking connection and contact with interactions to the extent that they are wanted or needed. A third aspect which is not very often used in discussions is inaction which refers to a state where individuals choose or are unable to take part in social action and for various reasons are disconnected from concepts of 'we-ness' and civic society. Social isolation and loneliness have been associated with lower reported life satisfaction, alcoholism, suicide and physical illness. Analyses of morbidity and mortality show stronger associations with social isolation than with loneliness.

There is still little research on the connections between music and participation with the attempt to reduce isolation. The interest in the positive effects and values of the connections between music and participation has however increased enormously over the last two decades. On a community level, there is a growing tendency to view music as a source for successful interventions for participation. On the individual level, music is revealed as an instrument in the repertoire of 'self-technologies', which aims at regulating our bodies, emotions and cognitive orientations. People use listening technology as a way to cope in their everyday lives. Some studies show that the music listening practices among youngsters, who are often the greatest music consumers, becomes health promoting when it helps them cope and regulate their feelings.

Part Three: Music as a Prophylactic Resource – Examples of Projects and Initiatives

A major challenge in the new field of music and public health is to transform knowledge from epidemiological studies and inspiration from theory on the relationship between music and health into practical projects for clinical as well as non-clinical groups, providing new treatment initiatives for the first and prophylaxis plans for the last. There is a lot of evidence to build upon, both from music therapy delivered by qualified music therapists (Kamioka et al. 2014; Li et al. 2015), from music medicine delivered by music therapists, nurses and doctors (Bonde 2015; Bradt, Dileo & Shim 2013) and music and health delivered or planned by a wide range of agents (Ruud 2013; Theorell and Kreutz 2012). In the final part of the anthology, illuminating examples of 'health musicking' are presented. They are all based on ongoing or finished empirical studies.

Chapter 10 (Söderström Gaden and Trondalen) explores ways in which musicking can support new parents and their children during the first year of life. It also examines the health promotion potential of a carefully designed music therapy programme. The chapter is based on a study at a public health clinic that was part of the Norwegian primary child services. Nine first-time mothers (and their infants) were invited to attend a weekly music therapy programme for 2 months. The group of participants was non-clinical, and the explorative, qualitative study was inspired by new, developmentally informed theory, as well as context-sensitive theory related to the development of a motherhood constellation in a contextual setting. Data included semi-structured interviews undertaken after the programme was concluded. The interviews were analyzed using interpretative phenomenological analysis (IPA), focusing on each participant's description of her experiences.

Results indicate that the mothers experienced positive development at a personal level, both within their relationships with their children and in relation to their everyday lives. The study also shows that the music therapy group at the public health clinic had the potential to increase mother-infant interaction and bonding and to support and empower first-time mothers in their new role. The emergence of a social network as a result of their shared participation in the programme was another result reported by the participants. The study documents how participation in a music therapy group can promote health in both mother and child. Supporting and strengthening parents during their child's first years of life using music therapy methods could be an important part of public health work in the future.

Chapter 11 (Balsnes) discusses the question 'How can choral singing promote public health?' Possible answers are based on analyses of choir members' stories about choral singing, drawn from various qualitative studies using participant observation and qualitative research interviews. The selected 'cases' are very different; the singers in question have varied backgrounds and current life situations, and the choirs they sing with are quite different with regard to purpose, level, repertoire and methods. Nevertheless, the singers explain that participation in choral singing has contributed to bettering their lives and thus improving their health.

The singers' stories are analyzed on the basis of the following questions: What is the meaning of the singing for the participants? What is it with choral singing that allows for a beneficial impact connected to health? What conditions must be present? And what characterizes the benefits that choral singing provides? The main objective of the chapter is the discussion of how choral singing can be a resource in public health work. Theoretical perspectives from music therapy and sociology of music illuminate the discussion, especially Ruud's categories concerning the relationship between music and quality of life and Stige's concept of *health musicking* (presented in Chap. 8).

Chapter 12 (Bojner Horwitz) takes as point of departure a sad fact that stress and burnout are common issues in the workplaces of our health-care institutions today – problems that must be addressed by organizational leaders. The chapter considers how the development of empathy can be one way of dealing with this complex issue by easing not simply the symptoms of stress and burnout but also addressing one of the causes of it.

Drawing on theories of mirroring, emotional brain and embodiment, the author examines how the combination of music and movement facilitates access to the emotions of others and helps people to better understand their experiences. It is also discussed how the training of empathy can be introduced into health care through programmes of cultural activities that facilitate an exploration of how the body can be used to increase awareness of one's own emotions and those of others. This work contributes to the growing body of evidence for how artistic programmes offer value to health care. The chapter shows how promotion of greater empathy as a practice of embodied leadership within the organization can contribute to the creation of a working environment that is 'humanized' through a culture of care.

Chapter 13 (Jacobsen, Lund and Bertelsen) describes the recent development within the field of arts and health in Denmark, zooming in on the use of music in a specific 'arts on prescription' model carried out in the Aalborg Municipality in an ongoing project (2017–2019). After a brief overview of established arts and health strategies in Scandinavia and the United Kingdom, the authors concentrate on a Danish project called 'culture vitamins' and its use of music and music interventions. Political issues, applicability and future research are discussed. In the chapter, the concept of 'music intervention' includes theory and practice from two different fields from clinical practice: 'music medicine' and 'music therapy'. 'Music medicine' is mainly a stimulus-response focused model whereas 'music therapy' is a psychodynamic humanistic model focused on interpersonal communication.

Based on guide experiences and statements from participants – citizens on long-term sick leave due to anxiety, stress or depression – it is illustrated how music listening in this specific framework, with the guidance of music therapists and the use of a specific music medicine application 'The Music Star' can be used to develop new coping strategies.

Chapter 14 (Batt-Rawden) presents and discusses a new approach to promoting health and quality of life in local contexts by teaching participants to use music as a 'technology' of health and self-care through the steps and actions of 'The Fellowship of Health Musicking Model'. The main purpose of this chapter is to increase knowledge as to how musical activities can promote mental and somatic health, hence to be considered important in public health matters. Secondly, how "The Fellowship of Musicking Model" may be used as a health promoting initiative. The *model* builds on a novel musical health promotion procedure developed by the author in 2007 as part of a PhD project. The aims of the study were to explore the role and significance of music in the life of men and women with long-term illnesses in or through different life phases, situations, events, issues and contexts. Secondly, to increase knowledge on how participants, through exposure to and exchange of new musical materials and practices, may learn to use music as a 'technology of self' in relation to health and healing.

The longitudinal study involved 9 men and 13 women, aged between 35 and 65 and was a pragmatic synthesis of elements of ethnography, grounded theory and action research. Eight in-depth interviews were conducted with each participant and open narratives were elicited from each of them, using a topic guide, two single CDs and four double compilations. Through involvement in this model, self-awareness and consciousness may be enhanced through the informal learning process, hence be adopted as a coping strategy independent of age, gender, diagnoses, illnesses and cultural differences. Health benefits from musicking may reduce stress, anxiety, depression and the need for medication, building coping capabilities, social inclusion and renewed strength. The model may contribute to a wider understanding of musical activities as a method or strategy in public health, health promotion and rehabilitation.

Some final words on the writing style in the book: We have allowed the authors to choose freely between writing in the first or in the third person. Some chapters cover similar or related theoretical material. We have inserted cross-references where relevant, in order to increase options of 'dialogue' where this could be relevant for the reader.

References

Bojner Horwitz, E., Hogstedt, C., Wistén, P., & Theorell, T. (Eds.). (2015). *Kultur & Folkhälsa: Antologi om forskning och praktik*. Stockholm: Tolvnitton Förlag.

Bonde, L. O. (2011). Health Musicing - Music Therapy or Music and Health? A model, empirical examples and personal reflections. *Music and Arts in Action, 3*(2), 120–140. Retrieved from http://musicandartsinaction.net/index.php/maia/article/view/healthmusicingmodel.

Bonde, L. O. (2015). Music and medicine and music therapy in hospitals – an overview of practices and evidence. *European Journal of Integrative Medicine, 7S*, 2. https://doi.org/10.1016/j.eujim.2015.09.012.

Bradt, J., Dileo, C. & Shim, M. (2013). Music interventions for preoperative anxiety. *Cochrane Database of Systematic Reviews 2013, Issue 6*. Art. No.: CD006908. https://doi.org/10.1002/14651858.CD006908.pub2.

Hanser, S. & Mandel, S. (2010). *Manage Your Stress and Pain Through Music*. Milwaukee: Berklee Press Publications.

Kamioka, H., Tsutani, K., Yamada, M., Park, H., Okuizumi, H., Tsuruoka, K., et al. (2014). Effectiveness of music therapy: A summary of systematic reviews based on randomized controlled trials of music interventions. *Patient Preference and Adherence, 8*, 727–754. https://doi.org/10.2147/PPA.S61340.

Li, H.-C., Wang, H.-H., Chou, F.-H., & Chen, K.-M. (2015). The effect of music therapy on cognitive functioning among older adults: A systematic review and meta-analysis. *Journal of the American Medical Directors Association, 16*(1), 71–77. https://doi.org/10.1016/j.jamda.2014.10.004.

Ruud, E. (2013). Can music serve as a "cultural immunogen"? An explorative study. *International Journal of Qualitative Studies on Health and Well-being, 8*(1), 20597. https://doi.org/10.3402/qhw.v8i0.20597

Stige, B. (2002). *Culture-Centered Music Therapy*. Gilsum NH: Barcelona Publishers.

Stige, B., & Aarø, L. E. (2012). *Invitation to community music therapy*. New York NY: Routledge.

Sifneos, P. E. (1973). The prevalence of 'alexithymic' characteristics in psychosomatic patients. *Psychother Psychosom*. *22*(2), 255–62.

Sifneos, P. E. (1996). Alexithymia: past and present. *American Journal of Psychiatry*, 153 (Suppl. 7), 137–14.

Theorell, T. (2014). *Psychological health effects of musical experiences: Theories, studies and reflections in music health science*. Dordrecht: Springer.

Theorell, T., & Kreutz, G. (2012). Epidemiological studies of the relationship between musical experiences and public health. In R. A. R. MacDonald, G. Kreutz, & L. Mitchell (Eds.), *Music, health, and wellbeing* (pp. 424–435). Oxford: Oxford University Press.

Part I
Epidemiological Research in Music and Public Health in the Nordic Countries

Chapter 2
Music and Health in Everyday Life in Denmark: Associations Between the Use of Music and Health-Related Outcomes in Adult Danes

Ola Ekholm and Lars Ole Bonde

Introduction

The relationship between music (active music making and receptive experiences), health and wellbeing has been explored extensively in the last 20 years by researchers from music psychology, music therapy, music sociology and medicine (MacDonald et al. 2012). A variety of scientific methods have been used to study how music 'works' as a health agent or 'cultural immunogen' (Ruud 2013), including interviews, narratives, observations, survey studies, brain scannings and randomized controlled trials (Theorell 2014). In this way, scientific evidence of music as a health resource is growing; however, epidemiological studies have been sparse, and especially longitudinal studies are warranted (MacDonald et al. 2012). The association between participation in cultural activities and self-rated health has been studied in countries like Sweden (Bygren et al. 2009; Konlaan 2001), Norway (Cuypers et al. 2012), Finland (Hyyppä and Mäki 2001) and Poland (Węziak-Białowolska and Białowolski 2016), and in all studies a small but robust association was found. However, as the authors of the recent Polish study write, no causal connection can be made in these cross-sectional studies, and differences between passive/receptive and active participation need to be taken more seriously into account (Węziak-Białowolska and Białowolski 2016).

The study presented here is the first Danish study on music and public health. It is a cross-sectional study, and therefore it has the same limitations as most of the studies mentioned above. However, the study has a clear and narrow focus on associations between *music* and health, not cultural participation in a broader sense.

O. Ekholm
National Institute of Public Health, University of Southern Denmark, Copenhagen, Denmark
e-mail: oek@niph.dk

L. O. Bonde (✉)
Department of Communication and Psychology, Aalborg University, Aalborg, Denmark
e-mail: lobo@hum.aau.dk

L. O. Bonde, T. Theorell (eds.), *Music and Public Health*,
https://doi.org/10.1007/978-3-319-76240-1_2

Therefore, the study reveals many associations between music and health that have not been addressed in this scale before. We have previously published some of the results (Ekholm et al. 2015, 2016; Bonde et al. 2018). In this chapter, after the presentation of a theoretical framework and the methodology, we will give an overview of the results, based on a discrimination between five distinct groups of informants: active and non-active amateur musicians, active and non-active professional musicians and non-musicians. Results include (a) an examination of associations between singing/playing and various health-related outcomes such as mental and physical health, pain and discomforts, obesity and other health behaviours; (b) the uses of music in everyday life; and (c) associations between attending musical live performances and beliefs in music as a health resource. A discussion will include the relevance of the findings for prophylactic and other practical music and health initiatives.

Theoretical Framework: Health Musicking

The field of 'music and health' covers a broad spectrum of activities and experiences, from lay therapeutic use of music in everyday life to the highly specialized procedures of music therapy (DeNora 2000; Trondalen and Bonde 2012). Bonde developed a 'map' of the field, inspired by the concept of 'health musicking' that was developed theoretically by Stige (Stige 2012; Stige and Aarø 2012). Stige was inspired by Small's concept of musicking ('Any activity involving or related to music performance, such as performing, listening, rehearsing, or composing'), (Small 1998, p. 9), and defined health musicking as

> The appraisal and appropriation of the health affordances of the arena, agenda, agent, activities, and artefacts of a music practice.

According to Batt-Rawden et al. (2007) *health musicking* contributes to promoting resilience, coping and recovery, while Bonde (2011) relates it to four main objectives: (1) the formation and development of identity, (2) the development of communities and values, (3) the lay and professional use of music and sound to support and help individuals and (4) the sharing and creation of musically designed environments. This is reflected in a quadrant model, where different arenas, agendas, agents, activities and artefacts are related to the four objectives (Fig. 2.1).

Most of the existing evidence – outside the specific field of clinical music therapy, including case studies as well as randomized controlled trials, (Theorell 2014; Koelsch 2013, 2015) – addresses the health benefits of choral singing. Clift et al. (2010) identified four specific factors to explain these benefits: (1) the experience of positive emotions, (2) focused attention, (3) deep breathing and (4) social support. Recent studies of choral singing as health promotion are summarized in Balsnes' chapter in the present book. The broader epidemiological studies mentioned above did not focus on choral singing or other musical activities; they gave a more general insight in associations between participation in cultural activities and health (Theorell and Kreutz 2012). Konlaan (2001) conducted a cohort study (ULF) in

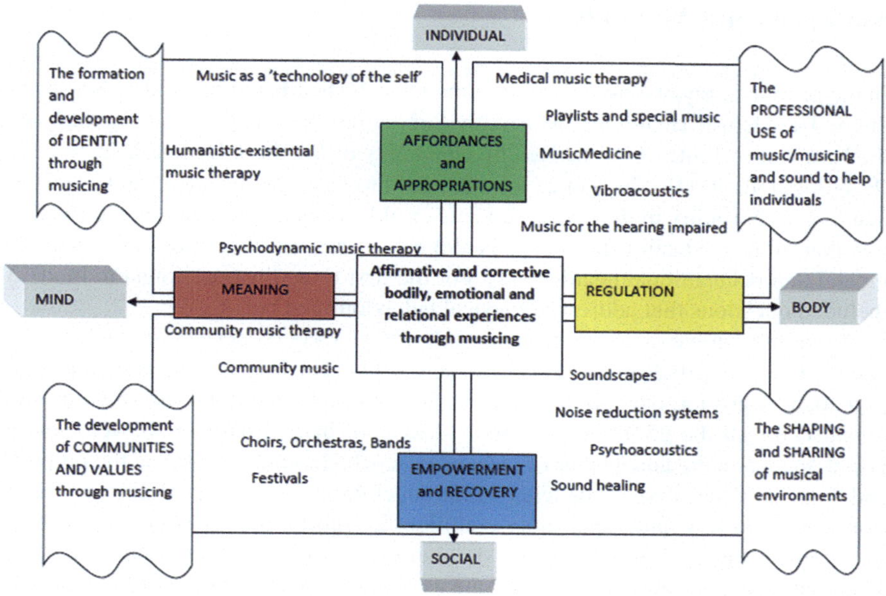

Fig. 2.1 Health musicking: a descriptive model

Sweden, where 10,609 men and women aged 25–74 were interviewed about their health as related to attendance of cultural activities. Participants were followed over 14 years (1982–1996) with regard to survival. The researchers found that there was a higher mortality risk for those informants who never or rarely visited the cinema, concerts, museums or art exhibitions compared to those visiting them more often. Regular visits to concerts significantly predicted survival per se, but no advantage of regular music making was found. A similar pattern can be found in the Norwegian HUNT study with more than 50,000 informants (Cuypers et al. 2012). Statistically significant, gender-specific associations were found between several receptive and creative cultural activities and self-rated health, depression, anxiety and satisfaction with life. Again, active participation was not found to be more beneficial than receptive experiences.

In a recent Polish study (Węziak-Białowolsk and Białowolski 2017), the positive association between cultural attendance and self-reported health was confirmed in a biennial longitudinal Polish household panel study with adult participants in four waves over 6 years. However, a positive *causative* relationship could not be confirmed, and no evidence was found to corroborate a positive impact from cultural attendance on physical health. Therefore, the authors conclude 'future research should investigate the causative influence of active participation in creative activities on health outcomes as, in contrast to passive attendance, it may be influential'. Even if the present study is cross-sectional and therefore only can establish associational connections between music and health, it does look more closely into active musicking, not only passive/receptive attendance.

Materials and Methods

In the present study, we used data from the Danish Health and Morbidity Survey in 2013. The Danish Health and Morbidity Survey has been carried out regularly by the National Institute of Public Health, University of Southern Denmark since 1987 (Ekholm et al. 2009). The aim of the surveys is to describe the status and trends in health and morbidity in the adult (16 years or older) Danish population and the factors that influence health status (e.g. health behaviour, social relations and environmental and occupational health risks). For the first time, the questionnaire in 2013 included questions that addressed musical behaviour and beliefs.

A random sample of 25,000 Danish adults (16 years or older) was drawn from the Danish Civil Registration System (Pedersen 2011). This register includes information such as gender, age, address and marital status. Background information was obtained for all the 25,000 invited individuals. All invited individuals were sent a postal questionnaire and a letter of introduction, which briefly described the purpose and content of the survey. The letter of introduction also emphasized that participation was voluntary and completely confidential. Individuals could choose either to complete the enclosed paper questionnaire or an identical web questionnaire. The invited individuals received a unique username and password in order to access the web questionnaire. In all, 14,265 individuals (participation rate: 57%) completed the questionnaire. As described in more detail elsewhere, non-response was significantly associated with gender, age and marital status (Ekholm et al. 2016).

All respondents were asked if they ever have voluntarily sung in a choir or sung or played a musical instrument in a band, orchestra or a musical ensemble. The possible answer categories were yes, sing or play now; yes, previously; and no. Those respondents who answered affirmatively were then asked: ‘Have you ever sung or played as a professional musician?’ Professional was defined as having played/sung for money. The possible answer categories were yes and no. Based on the answers to these two questions, the following five groups were created: active professional musicians, non-active professional musicians, active amateur musicians, non-active amateur musicians and non-musicians. The respondents were also asked if they use music for one or several of the following reasons in everyday life: for relaxation, to gain energy (e.g. to improve exercise performance), to get into a certain mood or to change mood, to express or explore feelings, to get to know yourself as a person, and as a means of concentration, not using music for anything special in everyday life (multiple choices were possible). The questionnaire also included the following question: ‘How often do you attend live musical performances?’ The possible answer categories were never/rarely, one to three times a year and more than three times a year. Furthermore, respondents were asked if they believe that music activities and music experiences can help to stay healthy. The possible answer categories were yes; definitely; yes, to some extent; and no; don’t know.

Health-related quality of life was assessed by using the physical component summary and the mental component summary scores, respectively, from the Short Form-12 (SF-12) version 2 (Ware et al. 1996). The summary scores were normed to the US population, and the tenth percentiles were used as cut-off points to define poor physical and mental health-related quality of life, respectively. Data on overall self-rated health were obtained from the SF-12 question: 'In general, would you say that your health is excellent, very good, good, fair or poor?' The Perceived Stress Scale (PSS) was used to measure the perception of stress (Cohen et al., 1983). A sum stress score was computed, with higher scores indicating greater stress. High perceived stress was defined as a PSS score at or above the 80% percentile (this threshold corresponds to a score of 18 or higher).

Symptoms, pain or complaints in general during the past 2 weeks were assessed with a checklist that included the following eight types of pain or discomforts: pain or discomfort in the shoulder or neck; pain or discomfort in the back or lower back; pain or discomfort in the arms, hands, legs, knees, hips or joints; fatigue; headache; sleeping problems or insomnia; melancholy, depression, and unhappiness; anxiety, nervousness, restlessness and apprehension. Health problems were also assessed with a standard checklist. The following conditions and diseases were included in the present study: asthma, diabetes, hypertension, osteoarthritis, rheumatoid arthritis, migraine or frequent headaches, spinal disc herniation and tinnitus. The answer categories were no, I have never had this; yes, I have this now; and yes, I have had this previously.

The respondents' alcohol habits were assessed in several ways in the present study. The following beverage-specific question was used to identify individuals with a high alcohol intake: 'How many alcoholic drinks do you typically have each day in a week?' The intake was measured in number of standard drinks, with one drink equalling approximately 12 g of pure alcohol. A high alcohol intake was defined in accordance with The Danish Health Authority's guidelines (men, >21 drinks/week; women, >14 drinks/week). Binge drinking was assessed by the question, 'How often do you have five or more drinks on one occasion?' with the five possible answer categories: daily or almost daily, weekly, monthly, less than monthly and never. The six-item CAGE-C questionnaire was used to screen for problem drinking (Zierau et al. 2005). Smoking behaviour was assessed by asking the respondent whether they smoked or not. Daily smokers were asked how much they smoked a day on average. Heavy smoking was defined as ≥15 cigarettes per day. The use of cannabis was assessed based on the recommendations by the European Monitoring Center for Drugs and Drug Addiction (European Monitoring Center for Drugs and Drug Addiction 2002). Thus, the following question was asked to all respondents, 'Have you ever tried cannabis?' with the four possible answer categories: no; yes, during the past month; yes, during the past year (but not during the past month); and yes, previously (but not during the past year). Leisure time physical activity was assessed by the following question, 'If you look back at the past year, what would you say best describes your leisure activities?' with the following four answer categories: heavy exercise and competitive sports regularly and several times a week; exercise or heavy gardening at least 4 h a week; walking, biking or other light exercise at least 4 h a week; and reading, watching TV or other sedentary activity. Respondents in the last category were classified as sedentary in leisure time. The validated dietary quality score was used to assess dietary habits

(Toft et al. 2007). Unhealthy diet was broadly defined as having a low intake of fruit, vegetables and fish and a high amount of saturated fat. Self-reported height and weight were used to calculate body mass index (BMI).

Data on cohabitation status were obtained by combining survey data on whether the respondent was living alone or with a partner and data on marital status from the Danish Civil Registration System. Socioeconomic status was measured by self-reported highest completed education level. Moreover, the Nordregio classification of urban and rural areas was used to divide the 98 Danish municipalities into four types of municipalities: urban, intermediate, rural and peripheral (Nordregio 2010).

Statistical Analysis

Descriptive statistics were presented as percentages. Multiple logistic regression models were used to investigate the associations between musicians and the various health-related outcomes. The results were presented as odds ratios (ORs) with 95% confidence intervals (CIs). The ORs were adjusted for gender, age, cohabitation status, highest completed education level and type of municipality. We also sought to examine the potential interaction between gender and musicians with regard to the various health-related outcomes. The analyses revealed a significant interaction between gender and musicians with regard to problematic drinking (no other significant interactions were observed).

Calibration weighting was applied in order to reduce the impact of non-response bias on the estimates (Särndal and Lundström 2005). The weights were computed by Statistics Denmark based on information from administrative registers (e.g. gender, age, educational level, income, employment status, marital status, country of origin and healthcare utilization) for all the invited subjects. Statistical analyses were performed using SAS version 9.3.

Results

Table 2.1 shows the characteristics of the five groups in the study population. In all, 207 (1.6%) and 503 (3.4%) respondents reported that they were active professional or amateur musicians, respectively. Furthermore, 203 (1.5%) and 3896 (28.0%) respondents reported that they previously have been professional or amateur musicians (i.e. non-active musicians), respectively. Active professional musicians were more likely to be men, young and higher educated. On the contrary, we found that active amateur musicians were more likely to be women and in an older age category.

Active professional musicians had 2.67 (95% CI: 1.30–5.48) times higher odds of reporting good (excellent, very good or good) health than non-musicians (i.e. individuals who never sung in a choir or sung or played a musical instrument in a band, orchestra or a musical ensemble) (Table 2.2). However, active professional

Table 2.1 Characteristics of the study population. Percentage

	Professional musicians		Amateur musicians			
	Active	Non-active	Active	Non-active	Non-musicians	All
Gender						
Men	70.7	54.8	38.5	33.6	56.1	49.4
Women	29.3	45.2	61.5	66.4	43.9	50.6
Age						
16–24 years	28.3	22.1	22.8	18.5	11.4	14.2
25–44 years	38.9	40.6	18.7	37.7	28.6	31.1
45–64 years	21.4	28.1	24.8	29.1	35.8	33.2
65–79 years	11.1	8.0	27.4	11.5	18.8	16.8
≥80 years	0.3	1.2	6.3	3.3	5.5	4.7
Cohabitation status						
Married	36.0	41.5	45.7	46.9	51.8	49.8
Cohabiting	20.9	16.5	8.2	14.9	13.2	13.7
Single (divorced, widowed or unmarried)	43.1	41.9	46.1	38.2	35.0	36.5
Combined school and vocational education						
<10 years	3.0	2.0	8.7	5.9	14.1	11.3
10–12 years	25.5	24.4	25.2	19.8	25.6	24.0
13–14 years	16.9	20.6	20.8	26.7	30.1	28.4
≥15 years	46.6	44.3	35.3	41.2	25.7	31.1
Other (in school, foreign education, etc.)	8.0	8.7	9.9	6.4	4.4	5.3
Type of municipality						
Urban	64.1	64.6	47.4	52.8	48.2	49.9
Intermediate	8.4	13.6	16.0	15.1	15.0	15.0
Rural	20.1	18.4	26.8	25.1	28.0	26.8
Peripheral	7.3	3.4	9.8	7.0	8.8	8.3
Number of respondents	207	203	503	3896	9120	13,929

musicians were more likely to report high perceived stress levels (although this result was on the borderline of statistical significance). Thus, they had 1.38 (95% CI: 1.00–1.92) times higher odds of having high levels of perceived stress than non-musicians. Interestingly, neither physical nor mental health was associated with musical behaviours.

Table 2.3 shows the associations between musical behaviours and very bothering pain and discomforts. Active professional musicians were more likely (OR: 1.81, 95% CI: 1.24–2.66) to have been very bothered by pain or discomfort in the back or lower back during the past 2 weeks than non-musicians. They were also more likely (OR: 1.82, 95% CI: 1.19–2.77) to have been very bothered by sleeping problems and insomnia during the past 2 weeks than non-musicians.

Table 2.2 Various health-related outcomes by musicians and non-musicians. Percentages and adjusted odds ratios[a] with 95% confidence intervals

		Professional musicians		Amateur musicians		Non-musicians	P-value[b]
		Active	Non-active	Active	Non-active		
Excellent, very good or good self-rated health	%	96.1	90.7	88.7	88.5	84.2	
	OR	**2.67 (1.30–5.48)**	1.30 (0.78–2.17)	1.25 (0.92–1.70)	1.10 (0.97–1.25)	1	0.025
Poor physical health (SF-12)	%	11.3	12.6	9.0	11.9	10.6	
	OR	1.16 (0.75–1.80)	1.07 (0.69–1.67)	0.84 (0.59–1.20)	1.04 (0.91–1.19)	1	0.762
Poor mental health (SF-12)	%	10.6	11.6	8.5	10.8	9.5	
	OR	1.19 (0.76–1.86)	1.14 (0.72–1.79)	0.90 (0.62–1.29)	1.04 (0.92–1.12)	1	0.831
High perceived stress (PSS)	%	25.4	18.6	19.6	23.3	22.6	
	OR	1.38 (1.00–1.92)	0.69 (0.47–1.03)	0.80 (0.62–1.04)	1.02 (0.90–1.10)	1	0.039

Note: Odds ratios in bold indicate statistical significant difference from the reference group
[a]ORs adjusted for gender, age, cohabitation status, combined school and vocational education and type of municipality
[b]The p-values indicate the statistical significance of the associations between the explanatory variable and the various health-related outcomes

No significant associations were observed between musical behaviours and various diseases and health problems, with one exception (Table 2.4). Both active and non-active professional musicians were more likely to have tinnitus than non-musicians. In addition, active amateur musicians also had significantly increased odds of having tinnitus.

Playing music (or singing) professionally was associated with health-risk behaviours such as binge drinking weekly and cannabis use in the last year (Table 2.5). Thus, active professional musicians had 1.98 (95% CI: 1.44–2.74) times higher odds of binge drinking weekly and 2.17 (95% CI: 1.59–2.96) times higher odds of being classified as problematic drinkers than non-musicians. In addition, non-active professional musicians had 2.00 (95% CI: 1.33–3.00) times higher odds of having a high alcohol intake and 1.87 (95% CI: 1.34–2.60) times higher odds of being classified as problematic drinkers than non-musicians.

Table 2.5 also shows that playing music (or singing) at an amateur level was generally associated with healthy behaviours (i.e. less likely to smoke tobacco, being classified as problematic drinkers, being sedentary in leisure time and having an unhealthy diet than non-musicians). However, one finding stands out: active amateur musicians were more likely to use cannabis than non-musicians. The analyses revealed a significant interaction between gender and musicians with regard to problematic drinking (data not shown). Thus, a high prevalence was observed among both active professional men (28.0%) and women (29.1%). However, large gender differences were observed among non-active professional musicians. In all,

Table 2.3 Very bothering pain or discomforts within the past 2 weeks by (active and non-active) professional and amateur musicians. Percentages and adjusted odds ratios[a] with 95% confidence intervals

Pain or discomfort		Professional musicians		Amateur musicians		Non-musicians	*P*-value[b]
		Active	Non-active	Active	Non-active		
Pain or discomfort in the shoulder or neck	%	12.8	12.4	14.0	15.3	14.0	
	OR	1.19 (0.78–1.81)	1.02 (0.66–1.57)	1.12 (0.84–1.49)	**1.13 (1.00–1.27)**	1	0.343
Pain or discomfort in the back or lower back	%	15.6	11.6	14.9	13.3	15.8	
	OR	**1.81 (1.24–2.66)**	0.98 (0.61–1.55)	1.00 (0.75–1.33)	0.94 (0.83–1.06)	1	0.027
Pain or discomfort in the arms, hands, legs, knees, hips or joints	%	9.9	11.0	16.0	13.4	14.4	
	OR	0.98 (0.63–1.55)	0.87 (0.54–1.40)	1.18 (0.89–1.56)	1.02 (0.90–1.15)	1	0.776
Fatigue	%	12.6	16.5	12.1	16.6	13.5	
	OR	1.11 (0.73–1.68)	1.21 (0.81–1.81)	0.90 (0.66–1.21)	**1.13 (1.01–1.27)**	1	0.193
Headache	%	6.2	5.4	5.3	8.1	7.0	
	OR	0.95 (0.53–1.73)	0.80 (0.43–1.48)	0.78 (0.50–1.22)	1.02 (0.87–1.19)	1	0.767
Sleeping problems or insomnia	%	12.1	8.9	7.1	10.8	9.6	
	OR	**1.82 (1.19–2.77)**	0.92 (0.54–1.57)	0.75 (0.51–1.10)	**1.20 (1.04–1.37)**	1	<0.001
Melancholy, depression, unhappiness	%	4.6	7.6	5.4	6.1	5.8	
	OR	0.87 (0.46–1.66)	0.91 (0.49–1.70)	0.90 (0.58–1.40)	0.98 (0.82–1.17)	1	0.977
Anxiety, nervousness, restlessness or apprehension	%	5.2	4.7	4.2	5.4	4.9	
	OR	1.38 (0.75–2.54)	0.58 (0.24–1.38)	0.91 (0.56–1.48)	1.10 (0.91–1.32)	1	0.426

Note: Odds ratios in bold indicate statistical significant difference from the reference group.
[a]ORs adjusted for gender, age, cohabitation status, combined school and vocational education and type of municipality
[b]The *p*-values indicate the statistical significance of the associations between the explanatory variable and the various health-related outcomes

34.8% among non-active professional men was classified as problematic drinkers. The corresponding prevalence among women was 14.2%.

Table 2.6 shows the association between age and use of music in everyday life. The use of music in everyday life varies considerably with age. The most frequent reason for using music in everyday life was to relax. In all, 65.2% of the population reported this as a reason for using music. Other frequently reported reasons were to get into a certain mood or to change mood (49.8%) and to gain energy (41.2%).

Table 2.4 Various self-reported diseases and health problems by (active and non-active) professional and amateur musicians. Percentages and adjusted odds ratios[a] with 95% confidence intervals

Pain or discomfort		Professional musicians		Amateur musicians		Non-musicians	*P*-value[b]
		Active	Non-active	Active	Non-active		
Asthma	%	5.2	4.2	6.0	6.5	5.7	
	OR	1.00 (0.53–1.87)	0.82 (0.40–1.69)	0.95 (0.60–1.50)	1.19 (1.00–1.42)	1	0.340
Diabetes	%	4.5	3.6	5.5	3.8	6.0	
	OR	1.60 (0.79–3.24)	1.18 (0.53–2.63)	0.97 (0.60–1.55)	0.96 (0.77–1.20)	1	0.720
Hypertension	%	9.1	12.0	20.2	13.0	19.3	
	OR	0.87 (0.52–1.45)	1.21 (0.74–1.98)	1.02 (0.77–1.36)	0.97 (0.85–1.10)	1	0.883
Osteoarthritis	%	9.2	10.6	22.9	14.7	20.0	
	OR	0.94 (0.56–1.57)	1.00 (0.60–1.65)	1.16 (0.88–1.52)	0.97 (0.86–1.10)	1	0.809
Rheumatoid arthritis	%	2.9	2,2	4.8	4.0	5.4	
	OR	1.11 (0.49–2.51)	0.72 (0.26–1.97)	0.94 (0.56–1.56)	1.04 (0.84–1.29)	1	0.951
Migraine or frequent headaches	%	8.9	10.0	10.0	14.2	11.7	
	OR	0.85 (0.52–1.41)	0.86 (0.53–1.40)	0.83 (0.59–1.17)	1.07 (0.94–1.21)	1	0.500
Spinal disc herniation	%	6.6	3.1	9.9	7.6	8.9	
	OR	1.25 (0.72–2.17)	0.48 (0.21–1.11)	1.21 (0.84–1.73)	1.09 (0.93–1.28)	1	0.220
Tinnitus	%	20.1	17.0	15.2	9.7	12.1	
	OR	**2.65 (1.85–3.80)**	**2.26 (1.51–3.38)**	**1.59 (1.18–2.12)**	1.13 (0.98–1.30)	1	<0.001

Note: Odds ratios in bold indicate statistical significant difference from the reference group.
[a]ORs adjusted for gender, age, cohabitation status, combined school and vocational education and type of municipality
[b]The *p*-values indicate the statistical significance of the associations between the explanatory variable and the various health-related outcomes

The analyses revealed only small gender differences (data not shown). Young adults were much more likely to use music in everyday life than elderly. Hence, 43.3% among individuals aged 80 or older did not use music for anything special in everyday life. The corresponding percentage among 16–24–year- olds was 3.6%.

In all, 14.7% of the adult Danish population reports that they attend live musical performances more than three times a year, and 38.2% report that they attend live performances one to three times a year. The percentage of individuals attending live musical performances decreases with age (Fig. 2.2). Our data revealed a strong association between frequency of attending live musical performances and a belief that music activities and music experiences can help staying healthy (Fig. 2.3). Thus, 23.0% among individuals who never or rarely attends live musical performances

Table 2.5 Health behaviour and obesity by (active and non-active) professional and amateur musicians. Percentages and adjusted odds ratios[a] with 95% confidence intervals

		Professional musicians		Amateur musicians		Non-musicians	*P*-value[b]
		Active	Non-active	Active	Non-active		
Daily smoker	%	19.3	18.6	12.7	15.7	18.3	
	OR	1.29 (0.90–1.85)	1.23 (0.90–1.85)	**0.72 (0.54–0.98)**	0.98 (0.88–1.09)	1	0.088
Heavy smoker	%	7.9	9.8	3.2	6.8	9.4	
	OR	1.04 (0.61–1.77)	1.25 (0.76–2.06)	**0.42 (0.24–0.72)**	**0.86 (0.73–0.99)**	1	0,007
High alcohol intake	%	13.3	15.4	9.1	8.1	7.9	
	OR	1.43 (0.92-2.22)	**2.00 (1.33–3.00)**	0.97 (0.69-1.38)	1.03 (0.88–1.19)	1	0.010
Binge drinking (weekly)	%	28.0	19.4	13.1	13.7	12.6	
	OR	**1.98 (1.44–2.74)**	1.26 (0.87–1.82)	0.90 (0.67–1.22)	0.95 (0.84–1.07)	1	<0.001
Problem alcohol use (CAGE-C)	%	28.3	25.4	12.7	15.0	15.7	
	OR	**2.17 (1.59–2.96)**	**1.87 (1.34–2.60)**	**0.69 (0.51–0.94)**	**0.97 (0.51–0.94)**	1	<0.001
Cannabis use in the last year	%	20.2	13.8	10.7	8.5	5.4	
	OR	**2.48 (1.68–3.66)**	1.44 (0.90-2.32)	**1.76 (1.22–2.52)**	1.13 (0.96–1.33)	1	<0.001
Sedentary in leisure time	%	11.1	10.4	9.2	13.3	18.0	
	OR	0.81 (0.53–1.25)	0.70 (0.44–1.12)	**0.52 (0.37–0.74)**	**0.85 (0.76–0.96)**	1	0.003
Unhealthy diet	%	10.2	13.9	8.1	9.6	16.6	
	OR	0.63 (0.40–1.00)	1.04 (0.68–1.59)	**0.55 (0.38–0.79)**	**0.72 (0.63–0.82)**	1	<0.001
Obesity (BMI≥30)	%	10.0	11.6	8.5	14.3	14.1	
	OR	0.92 (0.57–1.48)	1.17 (0.75–1.82)	0.72 (0.51–1.01)	**1.28 (1.13–1.43)**	1	0.009

Note: Odds ratios in bold indicate statistical significant difference from the reference group
[a]ORs adjusted for gender, age, cohabitation status, combined school and vocational education and type of municipality
[b]The *p*-values indicate the statistical significance of the associations between the explanatory variable and the various outcomes

definitely believed that music activities and music experiences can help staying healthy. The corresponding prevalence rates among those attending live musical performances one to three times a year or more than three times a year were substantially higher (34.5% and 53.5%, respectively).

Individuals attending live performances were more likely to report better health (Table 2.7). For example, those who attended live musical performances more than one to three times a year had 2.39 (95% CI: 1.98–2.89) times higher odds of reporting good (excellent, very good or good) health than those who never or rarely attend

Table 2.6 The use of music in everyday life according to age. Percentages

	16–24 years	25–44 years	45–64 years	65–79 years	≥80 years	All
For relaxation	85.2	69.6	62.1	52.5	41.5	65.2
To gain energy (e.g. to improve exercise performance)	72.2	55.4	30.3	18.4	11.6	41.2
To get into a certain mood or to change mood	74.1	60.4	42.0	31.6	24.8	49.8
To express or explore feelings	40.0	22.7	14.2	9.8	5.6	19.4
To get to know yourself as a person	19.0	10.8	7.3	6.4	5.6	9.9
As a means of concentration	46.1	26.3	15.3	10.1	7.2	21.9
Not using music for anything special in everyday life	3.6	12.7	23.9	31.6	43.3	19.7

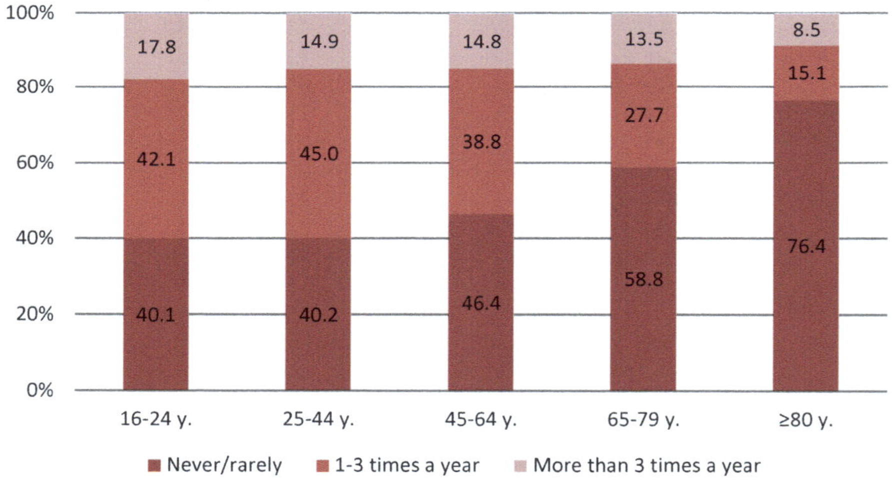

Fig. 2.2 Frequency of going to live musical performances according to age. Percentages

live musical performances. Those who attended live musical performances one to three times a year were also more likely to report good health than the reference group (OR: 1.92, 95% CI: 1.70–2.16). The possible interactions between frequency of attending live musical performances and active/non-active musicians in relation to the four health-related outcomes were not statistically significant (all *p*-values >0.05). This finding might reflect a lack of statistical power.

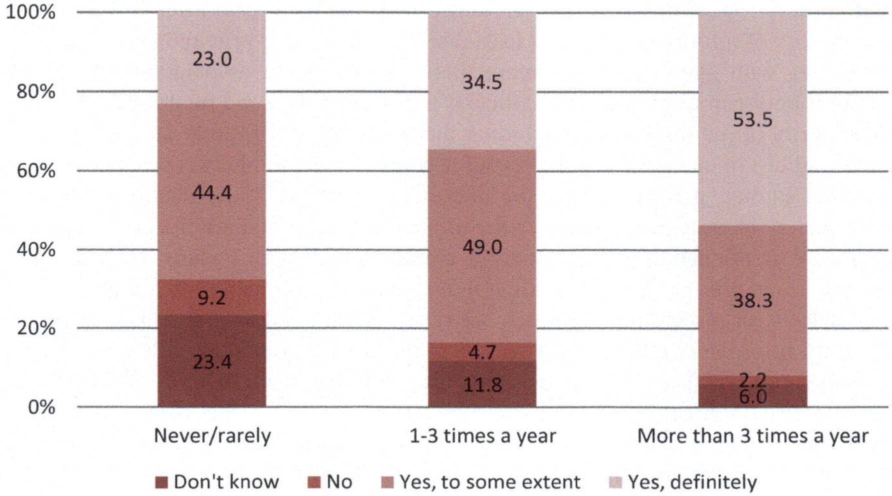

Fig. 2.3 Believe that music activities and music experiences can help to stay healthy according to frequency of attending live musical performances. Age-adjusted percentages

Table 2.7 Various health-related outcomes according to frequency of attending live musical performances. Percentages and adjusted odds ratios[a] with 95% confidence intervals

		More than three times a year	One to three times a year	Never/ rarely	*P*-value[b]
Excellent, very good or good self-rated health	%	92.4	90.8	79.9	
	OR	**2.39 (1.98–2.89)**	**1.92 (1.70–2.16)**	1	<0.001
Poor physical health (SF-12)	%	9.1	9.2	13.1	
	OR	**0.64 (0.53–0.77)**	**0.66 (0.58–0.75)**	1	<0.001
Poor mental health (SF-12)	%	8.5	8.1	12.0	
	OR	**0.66 (0.55–0.80)**	**0.63 (0.55–0.73)**	1	<0.001
Perceived stress (PSS)	%	17.0	19.7	27.1	
	OR	**0.58 (0.50–0.67)**	**0.69 (0.63–0.77)**	1	<0.001

Note: Odds ratios in bold indicate statistical significant difference from the reference group
[a]ORs adjusted for gender, age, cohabitation status, combined school and vocational education and type of municipality
[b]The p-values indicate the statistical significance of the associations between the explanatory variable and the various health-related outcomes

Discussion

The present study is based on a large national representative sample of adults, and for the first time in Denmark, associations between musical and health behaviour have been explored in a cross-sectional framework. The study documents that music is used in daily life by a majority of the informants – for many different purposes, with relaxation and mood/energy regulation as the most prominent (Table 2.6). A majority of informants attend at least one live concert per year and the prevalence decreases with age (Fig. 2.1). Interestingly, a significant association was documented between attending live concerts and reporting good health (Table 2.7). A majority of the informants, including those who never or rarely attend live concerts, believe in music as a health resource (Fig. 2.3). As in many comparable international studies, gender and age are important variables. These results are in line with the epidemiological studies of the influence of cultural participation on health that were mentioned in the introduction, e.g. the Swedish ULF study. However, the present study takes – for the first time in Denmark and probably internationally – a closer look on associations between active musicking and health. This is done by dividing informants into five groups based on their relationship with music: non-musicians (65.5%), active versus non-active amateur musicians (3.6/28.0%) and active versus non-active professional musicians (1.5/1.5%). The results show that active professional musicians suffer from various health problems and discomforts (e.g. sleeping problems, tinnitus) than non-musicians (Tables 2.3 and 2.4). In addition, active professional musicians were more likely to report a number of health risk behaviours (e.g. binge drinking, cannabis use) (Table 2.5). Paradoxically, active professional musicians report that they have excellent, very good or good self-rated health, even if they are at the same time were more likely to have high perceived stress than non-musicians (Table 2.2). Data are not detailed enough to allow interpretations of the influence of musical style or type of music activity on health behaviour or self-rated health (e.g. singing in a classical choir or playing in a rock band). However, in earlier American studies, certain specific musical genres and styles (rock, metal, electronic, hiphop, rap) have been identified as predictors of increased mortality (Bellis et al. 2012; Kenny et al. 2016).

All four types of musicians indicate that they have better self-reported health than non-musicians, and active amateur musicians tend to have more healthy behaviours than all other groups in all variables except cannabis use (Table 2.5).

Non-active amateur musicians seem to maintain some of these health benefits. However, they were more likely to be obese than non-musicians.

These results somehow contradict the cultural participation studies that found no extra benefit of active participation in cultural activities. The findings point towards a specific health potential of active musicking – especially for amateurs.

In the context of public health initiatives, the study highlights a number of issues that can be related to Stige's definition of *health musicking* (see Chap. 8 by Stige):

1. The gender and age variables suggest that, e.g. prevention and rehabilitation projects including musicking should be specifically aimed towards men over 45 and more general towards citizens over 65, who could benefit from both instructions in how to use music (arenas, artefacts) in everyday life as a health resource (agenda) and from sharing these experiences with other people. Musicians, music therapists and other professionals (agents) could design specific activities for non-clinical as well as clinical target groups, as exemplified and described in other chapters in this book.

2. Identification of the health benefits related to being an active amateur musician is a specific result of the present study. The famous Danish slogan 'Musik er godt. Spil selv' [Music is fine – play and sing!] from the 1960s could be reformulated in new campaigns as 'Musik er sundt. Spil og syng selv!' [Musicking is healthy – play and sing!]. Again, such campaigns could be specifically aimed at citizens in their 60s, also relating arenas and agendas to research establishing musicking as a potential means to prevent dementia (Balbag et al. 2014; Fung and Lehmberg 2016; see also the chapters by Ridder and Theorell & Ullén).
3. The present study has also shown an association between music in childhood/parental support and a later interest in music as a hobby or profession (not reported here, see Bonde et al. 2018). This association could be taken as point of departure (agenda) for music education initiatives, both in preschool institutions, public school and music schools. Learning children music as a health resource must start early and be designed in an age-specific way (Hallam 2010).
4. When looking at the use of music as a health resource in everyday life, age is an important variable. Young people use music more actively than mature and old people. This is probably not the result of formal education or training, more a 'lay craft' related to the omnipresence of music in public and private spheres and to the role of music among adolescents. Citizens over 50 could benefit from programmes (arenas, activities) promoting music listening, playing and singing as community agendas, both in working and in leisure time (see chapter by Ridder).
5. Attending live concerts (one to three times per year) is significantly associated with better self-reported health. From the present cross-sectional study, we cannot establish a causal link between attending concerts and improving health; however, when the result is related to more clinical knowledge about the benefits of interactive musicking (Stige et al. 2013), it is obvious that opportunities to experience live music in the community (arenas, agents) could be a simple, cost-effective health-promoting agenda for health service providers in many contexts.

Conclusion

This chapter confirms the associations between health and participation in music activities established in earlier Nordic epidemiological studies. It provides a new insight in the health benefits of daily musicking, especially for active amateur musicians, and the results, including the fact that gender and age are important variables, can be used as a knowledge platform to design *health musicking* programmes and initiatives. However, the most important public health measures are still about providing adequate family economy and education for all groups in society, ensuring safe communities without high risk of injuries and violence. It is therefore important that measures proposed above not will be in favour of the most basic public health measures. In line with this, public health initiatives including specially designed music making and receptive music experiences should be aimed especially at unprivileged groups in society who are not used to appropriate music as a health resource in daily life.

References

Balbag, M. A., Pedersen, N. L., & Gatz, M. (2014). Playing a musical instrument as a protective factor against dementia and cognitive impairment: A population-based twin study. *International Journal of Alzheimer's Disease*, 2014.

Batt-Rawden, K., Trythall, S., & DeNora, T. (2007). Health Musicking as cultural inclusion. In J. Edwards (Ed.), *Music: Promoting health and creating community in health care contexts* (pp. 64–82). Cambridge: Cambridge Scholars Press.

Bellis, M. A., Hughes, K., Sharples, O., Hennell, T., & Hardcastle, K. A. (2012). Dying to be famous: retrospective cohort study of rock and pop star mortality and its association with adverse childhood experiences. *BMJ Open*, *2*(6). https://doi.org/10.1136/bmjopen-2012-002089.

Bonde, L. O. (2011). Health Musicking – music therapy or music and health? A model, empirical examples and personal reflections. *Music and Arts in Action, 3*(2), 120–140. Retrieved from http://musicandartsinaction.net/index.php/maia/article/view/healthmusicingmodel.

Bonde, L.O., Juel, K., & Ekholm, O. (2018). Associations between music and health-related outcomes in adult non-musicians, amateur musicians and professional musicians – results from a nationwide Danish study. Nordic Journal of Music Therapy (in press).

Bygren, L. O., Weissglas, G., Wikstrom, B. M., Konlaan, B. B., Grjibovski, A., Karlsson, A. B., et al. (2009). Cultural participation and health: A randomized controlled trial among medical care staff. *Psychosomatic Medicine, 71*(4), 469–473. https://doi.org/10.1097/PSY.0b013e31819e47d4.

Cohen, S., Kamarck, T., & Mermelstein, R. (1983). A global measure of perceived stress. *Journal of health and social behavior, 24*, 385–396.

Clift, S., Hancox, G., Morrison, I., Hess, B., Kreutz, G., & Stewart, D. (2010). Choral singing and psychological wellbeing: Quantitative and qualitative findings from English choirs in a cross-national survey. *Journal of Applied Arts and Health, 1*(1), 19–34. https://doi.org/10.1386/jaah.1.1.19/1.

Cuypers, K., Krokstad, S., Lingaas Holmen, T., Skjei Knudtsen, M., Bygren, L. O., & Holmen, J. (2012). Patterns of receptive and creative cultural activities and their association with perceived health, anxiety, depression and satisfaction with life among adults: The HUNT study, Norway. *Journal of Epidemiology & Community Health, 66*, 698–703.

DeNora, T. (2000). *Music in everyday life*. Cambridge: Cambridge University Press.

Ekholm, O., Hesse, U., Davidsen, M., et al. (2009). The study design and characteristics of the Danish national health interview surveys. *Scandinavian journal of public health, 37*, 758–765.

Ekholm, O., Juel, K., & Bonde, L. O. (2015). Music and public health – an empirical study of the use of music in the daily life of adult Danes and the health implications of musical participation. *Arts & Health*, 1–15. https://doi.org/10.1080/17533015.2015.1048696.

Ekholm, O., Juel, K., & Bonde, L. O. (2016). Associations between daily musicking and health: Results from a nationwide survey in Denmark. *Scandinavian Journal of Public Health, 44*(7). https://doi.org/10.1177/1403494816664252.

European Monitoring Centre for Drugs and Drug Addiction. (2002). *Handbook for surveys on drug use among the general population*. Lisbon: EMCDDA.

Fung, C. V., & Lehmberg, L. J. (2016). *Music for life*. Oxford: Oxford University Press.

Hallam, S. (2010). The power of music: Its impact on the intellectual, social and personal development of children and young people. *International Journal of Music Education, 28*(3), 269–289. https://doi.org/10.1177/0255761410370658.

Hyyppä, M.T., & Mäki, J. (2001). Why do Swedish-speaking Finns have longer active life? An area for social capital research. *Health Promotion International, 16*, 196–198.

Kenny, D. T., Kenny, D. T., & Asher, A. (2016). Life expectancy and cause of death in popular musicians life expectancy and cause of death in popular musicians is the popular musician lifestyle the road to Ruin? *Medical Problems of Performing Artists, 31*(1), 37–44. https://doi.org/10.21091/mppa.2016.1007.

Koelsch, S. (2013). From social contact to social cohesion—the 7 Cs. *Music and Medicine, 5*(4), 204–209. https://doi.org/10.1177/1943862113508588.

Koelsch, S. (2015). Music-evoked emotions: Principles, brain correlates, and implications for therapy. *Annals of the New York Academy of Sciences, 13371*(1), 193–201. https://doi.org/10.1111/nyas.12684.

Konlaan, B. (2001). Cultural experience and health: the coherence of health and leisure time activities. Univ. of Umeå. Medical dissertations 2001.12.

MacDonald, R. A. R., Kreutz, G., & Mitchell, L. A. (2012). *Music, health, and well-being*. New York: Oxford University Press. http://search.ebscohost.com/login.aspx?direct=true&db=rih&AN=2012-00455&site=ehost-live.

Pedersen, C.B. (2011). The Danish Civil Registration System. *Scandinavian journal of public health, 39 (Suppl.7)*, 22–25. https://doi.org/10.1177/1403494810387965.

Ruud, E. (2013). Can music serve as a "cultural immunogen"? An explorative study. *International journal of qualitative studies on health and well-being, 8*(1), 20597. https://doi.org/10.3402/qhw.v8i0.20597.

Särndal, C.-E., & Lundström, S. (2005). *Estimation in surveys with non-response*. New York: Wiley.

Small, C. (1998). *Musicking*. Hanover: Wesleyan University Press.

Stige, B. (2012). *Health musicking: A perspective on music and health as action and performance. Music, health, and wellbeing*. https://doi.org/10.1093/acprof:oso/9780199586974.003.0014.

Stige, B., & Aarø, L. E. (2012). *Invitation to community music therapy*. New York: Routledge.

Stige, B., Ansdell, G., Elefant, C., Pavlicevic, M. (2013). Where music helps: Community music therapy in action and reflection. *Where music helps: Community music therapy in action and reflection*.

Theorell, T. (2014). *Psychological health effects of musical experiences: Theories, studies and reflections in music health science*. Dordrecht: Springer.

Theorell, T., & Kreutz, G. (2012). Epidemiological studies of the relationship between musical experiences and public health. In R. A. R. MacDonald, G. Kreutz, & L. Mitchell (Eds.), *Music, health, and wellbeing* (pp. 424–435). Oxford: Oxford University Press.

Trondalen, G., & Bonde, L. O. (2012). Music therapy: Models and interventions. In R. A. R. MacDonald, G. Kreutz, & L. Mitchell (Eds.), *Music, health, and wellbeing* (pp. 40–62). Oxford: Oxford University Press.

Węziak-Białowolska, D., & Białowolski, P. (2016). Cultural events – does attendance improve health? Evidence from a polish longitudinal study. *BMC Public Health, 16*.(August, 730. https://doi.org/10.1186/s12889-016-3433-y.

Zierau, F., Hardt, F., Henriksen, J. H., et al. (2005). Validation of a self-administered modified CAGE test (CAGE-C) in a somatic hospital ward: Comparison with biochemical markers. *Scandinavian Journal of Clinical and Laboratory Investigation, 65*, 615–622.

Chapter 3
Are Playing Instruments, Singing or Participating in Theatre Good for Population Health? Associations with Self-Rated Health and All-Cause Mortality in the HUNT3 Study (2006–2008), Norway

Bente Irene Løkken, Vegar Rangul, Dafna Merom, Ola Ekholm, Steinar Krokstad, and Erik R. Sund

Introduction

Participating in cultural activities may be conducive for health, and conversely, lack of participation may be health detrimental (Cuypers et al. 2011; Davies et al. 2014; Daykin et al. 2008; Gold et al. 2014; Thiel 2015). Numerous experimental studies

B. I. Løkken (✉) · E. R. Sund
Nord University, Levanger, Norway

HUNT Research Centre, Department of Public Health and Nursing, Faculty of Medicine, Norwegian University of Science and Technology (NTNU), Levanger, Norway
e-mail: bente.i.lokken@nord.no; erik.r.sund@ntnu.no

V. Rangul
Nord University, Levanger, Norway

HUNT Research Centre, Department of Public Health and Nursing, Faculty of Medicine, Norwegian University of Science and Technology (NTNU), Levanger, Norway

Norwegian Resource Centre for Arts and Health, Levanger, Norway
e-mail: vegar.rangul@ntnu.no

D. Merom
School of Science and Health, Western Sydney University, Sydney, NSW, Australia
e-mail: D.Merom@westernsydney.edu.au

O. Ekholm
National Institute of Public Health, University of Southern Denmark, Copenhagen, Denmark
e-mail: oek@niph.dk

S. Krokstad
HUNT Research Centre, Department of Public Health and Nursing, Faculty of Medicine, Norwegian University of Science and Technology (NTNU), Levanger, Norway

Norwegian Resource Centre for Arts and Health, Levanger, Norway
e-mail: steinar.krokstad@ntnu.no

L. O. Bonde, T. Theorell (eds.), *Music and Public Health*,
https://doi.org/10.1007/978-3-319-76240-1_3

have indicated that engagement in art-based activities (e.g. music, singing, theatre, painting, community cultural festival) has a therapeutic effect among clinical populations. Two systematic reviews of music-based interventions in health-care setting involving patients (Llovet 2017) or in aged care facilities for demented clients (Raglio et al. 2014) generally indicate that such interventions were associated with improved health outcomes, higher quality of life (QOL) and better cognitive, emotional and social outcomes. Music is therefore emphasized as important in the field of managing dementia and also considered as a tool for bringing back memory, eliciting conversation, calming down behaviour disturbances and improving moods (Raglio et al. 2014). In mental health care, music is an effective therapy for serious mental disorders, with slightly improvements to be seen after a few therapy sessions (Gold et al. 2009). However, cultural activities as such are practised by many members of the society as part of everyday life, but there is limited evidence on the public health significance of such activities. As good health is one of the key qualities of life (Aartsen et al. 2017), the question is whether participation in music, singing and theatre as a way of life can promote the health of the population as a whole.

The potential mechanisms between active participation in cultural activities and health are several. In a review by Cuypers et al., creative activities were recommended as part of patients' therapy, for example, music therapy has previously been emphasized in Norway and Sweden (Cuypers et al. 2011). It is hypothesized that such activities may affect biological processes in the human body (Theorell et al. 2007) and hence can be beneficial for the health of patients (Llovet 2017; Theorell et al. 2007; Thiel 2015). The mechanism by which cultural activities may improve the health of populations is assumed complex involving mental, social and biological possible pathways, implying cognitive, affective and behaviour pathways (Davies et al. 2016b; Holt-Lunstad et al. 2010). For example, social relationships may directly or indirectly encourage healthy behaviours, and being part of a social network gives individuals meaningful roles that provide self-esteem and purpose to life (Holt-Lunstad et al. 2010). Diba et al. performed a qualitative study with the aim to analyse the relationship of an experience in community theatre with the promotion of health. It clearly showed the importance of theatrical joint experience for the promotion of health by transforming the quality of relationships between people (Diba and D'Oliveira 2015). The quality and quantity of individuals' social relationships have been linked to physiological health and mortality (Holt-Lunstad et al. 2010). A meta-analytic review with participants from Europe, North America, Asia and Australia found an increased likelihood of survival with influence of social relationship, related to large networks, high contact frequency and higher levels of received support (Holt-Lunstad et al. 2010).

The social – as well as the physical – environment has effect on the body and the brain through the neuroendocrine, autonomic and immune system (McEwen 2012). Stress is a condition of mind-body interaction (McEwen 2006, 2012). Grape et al. investigated the effect of singing on the well-being of professional and amateur singers. The findings were that both groups had increased oxytocin level after singing lessons and felt more energetic and relaxed. After the singing performance, the amateurs reported increasing levels of joy and elatedness, whereas professionals did not. Singing among amateurs seemed to promote a sense of well-being and less arousal (i.e. stress) compared to professionals (Grape et al. 2002).

Although understanding the mechanism is important to support causal association between participation in cultural activity as a way of life and health outcomes, empirical evidence from population-based sample is necessary to establish the significance of cultural activity in the public health context. A significant association between cultural activity and self-reported health (SRH) has been observed in several cross-sectional studies with representative population samples (Cuypers et al. 2012; Ekholm et al. 2016; Johansson et al. 2001; Nenonen et al. 2014; Nummela et al. 2008; Wilkinson et al. 2007). In a study by Nenonen et al., the findings were that among other cultural activities, the music activity was evidently positively associated with SRH and quality of life (Nenonen et al. 2014). Both musically active men and women were more likely to report good SRH than individuals that were not active musically. Wilkinson et al. found among US residents that the more cultural activities people reported to attend, the better was their SRH (Wilkinson et al. 2007). Ekholm et al. investigated musically active individuals, playing instrument or singing at least 1 h every day, in association with SRH. Both musically active men and women were more likely to report good health (Ekholm et al. 2016). A study by Cuypers et al. conducted in Norway based on data from the third Nord-Trøndelag Health study (HUNT3) analysed data on both receptive and creative cultural activities. Findings indicate that people, who were culturally active experienced better SRH, were more satisfied with their lives and reported less symptoms of anxiety and depression (Cuypers et al. 2012). A similar study performed on adolescents, 13–19 years, found an association with participation and good health, good life satisfaction and good self-esteem (Hansen et al. 2015). However, cross-sectional studies cannot support causal association due to assessments of cultural activity and SRH at the same time point. It is possible that good health can make people actively participate in various cultural activities and, not conversely, that participation in cultural activities improves health.

There are a limited number of longitudinal cohort studies that have explored the associations between cultural activities and health outcomes. Johansson et al. investigated changes in SRH over 8 years in relation to the habit of attending cultural events in the community. Those who reduced their participation in cultural activities and those who were inactive at both occasions had a 65% higher risk to perceived their health as poor. An interesting finding was that among those who increased their cultural activity from being less active to being more active had about the same level of perceived good health as those who were active in both measurements, suggesting benefits can accrue even if started later in life (Johansson et al. 2001).

Other longitudinal studies focus on mortality as the main health outcome (Bygren 1996; Hyyppä et al. 2006; Konlaan et al. 2000; Merom et al. 2016; Väänänen et al. 2009; Windsor 2005). In a Swedish study by Bygren et al., the investigators found a significant association between participation in cultural activities, reading books or periodicals and making music or singing in a choir, and life expectancy (Bygren 1996). The same cohort was investigated for additional 14 years of follow-up, and similarly a higher mortality risk was observed among those who rarely participated compared to those who participate frequently (Konlaan et al. 2000). Among a working population in Finland, cultural engagement outside work life increased overall survival (Väänänen et al. 2009). In the Nurses' Health Study in US women who attended religious service more than once a week were 33% less likely to experience premature death compared to those who never attended religious service (Li et al. 2016). A longitudinal study from the US followed a small cohort with 469 subjects of volunteered healthy older adults, age above 75 years, the association with variety of leisure activities were examined in relation to cognitive decline and development of dementia. Among all activities, playing musical instrument or dancing were associated with about 70% reduction in the risk of developing dementia (Verghese et al. 2003). A prospective study of 6000 working Swedish men and women showed that those who reported that there were cultural activities organized in their worksites were more likely than others to report a favourable development of emotional exhaustion (proxy for burnout) after 2 years. This finding remained significant after adjustment for age, sex, education, emotional exhaustion at start and working conditions (Theorell et al. 2013). Clearly, there is a need for more prospective population-based studies to confirm possible health benefits of participation in arts and culture. The aim of this study was to explore whether people who actively engage in music, singing and theatre, have better self-rated health (SRH) and most importantly, whether participants in these activities on a regular basis survive longer compared to those who do not participate in these activities at all. Secondarily, we wanted to explore possible gender differences.

Material and Methods

Data were obtained from The Nord-Trøndelag Health Study (HUNT), which is a large Norwegian population-based cohort study with three consecutive waves of data collection conducted in 1984–1986 (HUNT1), 1995–1997 (HUNT2) and 2006–2008 (HUNT3). In all three surveys, all residents aged ≥20 years were invited to participate (Krokstad et al. 2013). The HUNT study collected data using comprehensive questionnaires, clinical measurements and blood and urine samples. The present study uses data from the third survey, HUNT3, where 94,120 adults were invited and in total 50,807 participants, 54.1%, participated (Krokstad et al. 2013). Out of these, 41,198 returned the questionnaires with information about cultural activities.

In the HUNT3 survey, the first questionnaire (Q1) was mailed together with the invitation and was delivered when meeting up at the examination station. At the examination station, questionnaire 2 (Q2) and 3 were handed out based on selection criteria such as age, gender and answers from Q1 and should be completed at home and returned to the HUNT Research Centre. The Nord-Trøndelag county has a homogenous population with a demographical and geographical and occupational structure fairly representative for the whole Norway (Holmen et al. 2003), though the region is lacking large cities.

Cultural Activities

The Q2 included the exposure to leisure time cultural activities that are divided into two questions: receptive and creative activities. In this article, we used the question about creative activities: *How many times in the last 6 months, have you participated in?* The response alternatives in the questionnaires were given: An association or club meeting/activity? Music, singing or theatre? Parish work? Outdoor activities? Dance? Sports or exercise? Each creative activity had response alternatives: 'more than 1x/week', '1x/week', '1–3x/month', '1–5x/6 months' and 'never'. The exposure variable is "How many times in the last 6 months have you participated in the following: Music, singing and theatre? From the five intervals, we derived a new variable with three categories: regular, occasional and never. 'Occasional': '1–3x/month' and '1–5x/6 months' and 'regular': '1x/week' and 'more than 1x/week'. For the other creative cultural activities, the response alternative was dichotomized into 'never' if the response was 'never' and 'ever' if the response was '1–3x/month', '1–5x/6 months', '1x/week' and 'more than 1x/week'.

A validation of the questions of cultural activities showed that it is a rough estimate of the activities in the population, and the questions are sufficiently precise to use at the group level in the population (Holmen et al. 2016).

Outcome Variable Self-Rated Health (SRH): Participants in the survey were asked to rate their health with the question: *How is your health at the moment?* The possible response alternatives were 'very good', 'good', 'not so good' and 'poor'. These four categories were collapsed into two categories creating a dichotomized variable: those who perceived their health as 'not so good or poor' versus those who perceived their health as 'good or very good'.

Outcome Variable All-Cause Mortality: Mortality data were obtained from the Norwegian Cause of Death Registry, administrated by the Norwegian Institute of Public Health. All data available in the registry are based on death certificates reported by doctors in accordance with the *International Classification of Diseases (ICD)*. Linkage was performed with Norwegian 11-digit personal identification number (Krokstad et al. 2012). All-cause mortality data were obtained from participation date until 31 December 2015.

Covariates: Gender and age data were obtained from the National Population Registry in Norway, containing information concerning everyone who either is or had been a resident in Norway (HUNT Research Centre 2017a). Age at participation date was used as a continuous variable. Education data were retrieved from Statistics Norway based on the highest achieved education measured as level of education completed on the 2007 census. The seven levels of *ISCED 1997* were collapsed into three main levels: primary (primary and lower secondary school), secondary (upper secondary and post-secondary school) or tertiary (first and second stage of tertiary education). Education level was used as a proxy measure of socioeconomic status. Marital status was divided into a dichotomous variable with the categories: being in a relationship or married vs. others (unmarried, separated, divorced or widow(er)). In addition, longstanding illness: *Do you suffer from longstanding (at least 1 year) illness or injury of a physical or psychological nature that impairs your functioning in your daily life?* Response alternatives were yes or no.

Ethics

The regional committee for Medical and Health Research Ethics (REK) in Norway approved this study (4.2006.250 REK-Midt), and participating was voluntary. Informed written consent was obtained prior to the participation at the screening station. The participants were informed in written form that they could withdraw the HUNT study at any time. In the consent, the participants approved for link data from national registers (HUNT Research Centre 2017b; Langhammer et al. 2012).

Statistical Analysis

Logistic regression models were used in cross-sectional analyses to study associations between creative participating in music, singing and theatre and SRH. For the logistic regression analysis, we report odds ratio (OR) and 95% confidence interval (CI). Cox proportional hazard regression models were used to study associations between creative participating in music, singing and theatre on all-cause mortality, whereas logistic regression was used to examine the relationships between creative participating in music, singing and theatre, and SRH. All analyses were done separately for men and women. We discarded 7731 (18.8%) participants with missing data and performed a complete case analysis for the remainders ($N = 33{,}467$).

Models were built sequentially whereby the first model was adjusted for age, model 2 was further adjusted for education level and marital status, model 3 was further adjusted for longstanding illness and model 4 was further adjusted for all the other creative activities: 'association or club meeting/activity', 'parish work', 'outdoor activities', 'dance' and 'work out or sports'.

Person-time was accrued from baseline participation date until the date of death, loss to follow-up or 31 December 2015, whichever came first. For the cox regression models, we report hazard ratios (HR), and 95% CI. IBM SPSS version 24 was used for all analysis.

Results

In total, 33,467 participants with completed answers on all variables are included in the analysis (Table 3.1). Data from 15,128 males, (45.2%) and 18,339 females (54.8%) age 19–97 (mean 52.7 SD 15.4) were analysed. In total 3038 individuals participated in music, singing and theatre on regular basis – slightly higher frequency among men (9.7%) than among women (8.5%). In addition, 9.5% of men and 12.1% of women occasionally participated in music, singing and theatre activities, and the rest reported that they had never participated in such activities during the last 6 months (80.7% of men and 79.4% of women). The socio-demographic and health characteristics, separately for men and women, by level of participation, are presented in Table 3.1.

Table 3.1 The proportion of cultural attendance in the different independent variables

	Music, singing and theatre								
	Men					Women			
	Total	Regular ($n = 1473$ (9.7))	Occasional ($n = 1443$ (9.5))	Never ($n = 12{,}212$ (80.7))	*P*-value	Regular ($n = 1565$ (8.5))	Occasional ($n = 2216$ (12.1))	Never ($n = 14{,}558$ (79.4))	*P*-value
Age					0.003				<0.001
<50	14,263	543	611	4636		965	1138	6370	
	(42.6)	(9.4)	(10.6)	(80.1)		(11.4)	(13.4)	(75.2)	
≥50	19,204	930	832	7576		600	1078	8188	
	(57.4)	(10.0)	(8.9)	(81.1)		(6.1)	(10.9)	(83.0)	
Education; level					<0.001				<0.001
Primary	6298	144	167	2290		159	270	3268	
	(18.8)	(5.5)	(6.4)	(88.0)		(4.3)	(7.3)	(88.4)	
Secondary	17,100	789	760	7112		628	935	6876	
	(51.1)	(9.1)	(8.8)	(82.1)		(7.4)	(11.1)	(81.5)	
Tertiary	10,069	540	516	2810		778	1011	4414	
	(30.1)	(14.0)	(13.3)	(72.7)		(12.5)	(16.3)	(71.2)	
Marital status					0.015				0.002
Marriage/relationship	20,262	984	953	7767		848	1326	8384	
	(60.5)	(10.1)	(9.8)	(80.0)		(8.0)	(12.6)	(79.4)	
Other	13,205	489	490	4445		717	890	6174	
	(39.5)	(9.0)	(9.0)	(82.0)		(9.2)	(11.4)	(79.3)	
Longstanding illness (>1 year)					0.003				<0.001
Yes	13,310	547	536	4951		512	814	5950	
	(39.8)	(9.1)	(8.9)	(82.1)		(7.0)	(11.2)	(81.8)	

(continued)

Table 3.1 (continued)

	Music, singing and theatre								
		Men				Women			
	Total	Regular (*n* = 1473 (9.7))	Occasional (*n* = 1443 (9.5))	Never (*n* = 12,212 (80.7))	*P*-value	Regular (*n* = 1565 (8.5))	Occasional (*n* = 2216 (12.1))	Never (*n* = 14,558 (79.4))	*P*-value
No	20,157	926	907	7261		1053	1402	8608	
	(60.2)	(10.2)	(10.0)	(79.8)		(9.5)	(12.7)	(77.8)	
Self-reported health					<0.001				<0.001
Poor	8501	306	276	2942		317	490	4170	
	(25.4)	(8.7)	(7.8)	(83.5)		(6.4)	(9.8)	(83.8)	
Good	24,966	1167	1167	9270		1248	1726	10,388	
	(74.6)	(10.1)	(10.1)	(79.8)		(9.3)	(12.9)	(77.7)	
Club member/organization work					<0.001				<0.001
Participating	14,264	918	919	4607		917	1417	5486	
	(42.6)	(14.2)	(14.3)	(71.5)		(11.7)	(18.1)	(70.2)	
Never	19,203	555	524	7605		648	799	9072	
	(57.4)	(6.4)	(6.0)	(87.6)		(6.2)	(7.6)	(86.2)	
Dance					<0.001				<0.001
Participating	12,459	780	754	3575		964	1344	5042	
	(37.2)	(15.3)	(14.8)	(70.0)		(13.1)	(18.3)	(68.6)	
Never	21,008	693	689	8637		601	872	9516	
	(62.8)	(6.9)	(6.9)	(86.2)		(5.5)	(7.9)	(86.6)	
Parish work					<0.001				<0.001
Participating	1796	168	184	330		202	324	588	
	(5.4)	(24.6)	(27.0)	(48.4)		(18.1)	(29.1)	(52.8)	
Never	31,671	1305	1259	11,882		1363	1892	13,970	
	(94.6)	(9.0)	(8.7)	(82.3)		(7.9)	(11.0)	(81.1)	

Outdoor activities					<0.001				<0.001
Participating	27,147	1348	1361	10,029		1402	2043	10,964	
	(81.7)	(10.6)	(10.7)	(78.7)		(9.7)	(14.2)	(76.1)	
Never	6061	118	72	2120		143	156	3452	
	(18.3)	(5.1)	(3.1)	(91.8)		(3.8)	(4.2)	(92.0)	
Sports					<0.001				<0.001
Participating	20,427	990	1069	7163		1175	1652	8378	
	(61.0)	(10.7)	(11.6)	(77.7)		(10.5)	(14.7)	(74.8)	
Never	13,040	483	374	5049		390	564	6180	
	(39.0)	(8.2)	(6.3)	(85.5)		(5.5)	(7.9)	(86.6)	

n = 33,467 (N = 41,198, missing 7731, 18.8%). P-value, Pearson chi-square. Values are expressed as number (%)

Younger women reported higher level of participation in music, singing and theatre than older women. There was no age difference by level of participation among men. The frequency of participation by education level varied in a gradient manner: individuals with higher level of education had greater level of regular or occasional participation in music, singing and theatre, and this was true for both men and women. The participation level among married/cohabitating couples was not different from the level among those not married, for both women and men.

Further, Table 3.1 presents the distribution of other leisure activities such as club membership, dancing, parish work, outdoor activities and sports. The distribution pattern across level of participation suggests that those who never participated in music, singing and theatre were more likely to never participate in other leisure activities, and this pattern was true for men and women. For example, the proportion of sports participants among regular participants in music singing and theatre was higher (21.3% in men and 25.2% in women) than among those who never participate in sports (14.3% in men and 13.4% in women) suggesting that people who engage in active leisure also tend to show greater participation in music, singing and theatre. In all, 25.4% of the 33,467 participants reported poor health at baseline. Those with poorer SRH were more likely to never participate in music, singing or theatre than those reporting good health; the differences were greater in women 83.8% vs. 77.7% than in men 83.5% vs. 79.9%.

Table 3.2 provides the estimates for the association between participation in music, singing and theatre and SRH. The age-adjusted model shows that participants who never attended music, singing and theatre activity in the last 6 month were 26% more likely (OR = 1.26; 95% CI: 1.15,1.39) to report poor SRH compared to participants that regularly participated in these activities. For women the estimated OR was 1.29 (95% CI: 1.13,1.48), and for men it was 1.19 (95% CI: 1.04,1.36). Further adjustment for the confounders gender, education and marital status (model 2) was made; the effect was attenuated for the whole population OR = 1.10 (95% CI: 1.00,1.21) but remained significant. Stratification by gender showed that the effect of never participating in music, singing and theatre remained significant for women OR = 1.16 (95% CI: 1.02,1.33) but not for men.

In model 3 we further adjusted for longstanding illness, and the results remained significant only for women OR = 1.17 (95% CI: 1.01,1.36). Adjustments for other creative activities (model 4) completely attenuated the association between music, singing and theatre on the SRH of women.

During 7 years of follow-up, in total 897 women and 1205 men died. That is in total 2102 deaths over an average follow-up of 270,258 person-years. In all, 86.5% out of those who died during this period never participated in music, singing and theatre; this proportion was similar for men and women. As expected, mortality rate was higher among the older population. Thus, out of 19,204 participants aged 50 years or above 2020 (10.5%), individuals died in the follow-up period, a death rate of 105 per 1000 people. By contrast, out of the 14,263 younger participants, 82 died (0.6%) in the follow-up period, a rate of 5.7 per 1000 people.

After adjusting for age (Table 3.3), those who never attended music, singing and theatre had 24% higher risk (95% CI: 1.03,1.50) of dying than those who regularly

Table 3.2 Odds ratio (OR) of poor self-rated health associated with participation in music, singing and theatre (n = 33,467)

Music, singing and theatre	Model 1			Model 2			Model 3			Model 4		
	All	Men	Women	*All*	Men	Women	*All*	Men	Women	*All*	Men	Women
Regular (once or more pr. wk.), ref.	1	1	1	1	1	1	1	1	1	1	1	1
Occasional	0.97	0.91	0.98	0.93	0.89	0.97	0.89	0.85	0.95	0.94	0.90	0.98
(1–3 times pr. month and 1–6 times pr. 6 months)	(0.86,1.10)	(0.75,1.09)	(0.83,1.15)	(0.82,1.05)	(0.74,1.07)	(0.82,1.15)	(0.78,1.03)	(0.69,1.05)	(0.79,1.14)	(0.81,1.07)	(0.72,1.11)	(0.81,1.18)
Never	1.26	1.19	1.29	1.10	1.05	1.16	1.10	1.03	1.17	0.98	0.92	1.06
	(1.15,1.39)	(1.04,1.36)	(1.13,1.48)	(1.00,1.21)	(0.91,1.20)	(1.02,1.33)	(0.98,1.22)	(0.88,1.20)	(1.01.1.36)	(0.88,1.10)	(0.78,1.08)	(0.90,1.23)

Model 1: adjusted for age
Model 2: also adjusted for gender, education level and marital status
Model 3: also adjusted for longstanding illness
Model 4: also adjusted for other creative cultural activities: 'association or club meeting/activity', 'parish work', 'outdoor activities', 'dance' and 'work out or sports'

Table 3.3 Association between participating in music, singing and theatre and all-cause mortality (n = 33,467)

Music, singing and theatre	Deaths/ person years	Model 1			Model 2			Model 3			Model 4		
		All	Men	Women	All	Men	Women	All	Men	Women	All	Men	Women
Regular (once or more pr. wk.), ref.	120/24,849	1	1	1	1	1	1	1	1	1	1	1	1
Occasional (1–3 times pr. month and 1–6 times pr. 6 month)	163/29,770	0.89	1.07	0.79	0.98	1.07	0.80	0.96	1.03	0.81	0.97	1.05	0.82
		(0.71, 1.13)	(0.78, 1.46)	(0.55, 1.15)	(0.77, 1.24)	(0.79, 1.45)	(0.55, 1.17)	(0.76, 1.22)	(0.76, 1.41)	(0.55, 1.18)	(0.77, 1.23)	(0.77, 1.43)	(0.56, 1.2)
Never	1819/215,638	1.24	1.48	1.07	1.27	1.42	1.03	1.26	1.40	1.02	1.09	1.23	0.87
		(1.03, 1.50)	(1.18, 1.87)	(0.78, 1.45)	(1.06, 1.53)	(1.12, 1.79)	(0.75, 1.4)	(1.04, 1.51)	(1.11, 1.76)	(0.75, 1.39)	(0.90, 1.32)	(0.97, 1.57)	(0.63, 1.2)

Model 1: adjusted for age
Model 2: also adjusted for gender, education level and marital status
Model 3: also adjusted for longstanding illness
Model 4: also adjusted for other creative cultural activities: 'Association or club meeting/activity', 'parish work', 'outdoor activities', 'dance' and 'work out or sports'

participated in these activities. Examining the association separately for men and women suggests that music, singing and theatre were significantly associated with longer survival among men. The HR of mortality for non-participating men was 1.48 (95% CI: 1.18,1.87) higher than participating men. This association remained strong and significant after adjusting for education, marital status and longstanding illness 1.40 (95% CI: 1.11, 1.76). The association was attenuated to 1.26 after adjusting for all other leisure participation, but the 95% confidence interval ranged from 0.97 to 1.57, indicating that there is still a possibility that the association is due to chance. By contrast, there was no association between music, singing and theatre and mortality among women in any of the models.

Discussion

The main findings from this Norwegian population-based study were that participation in music, singing and theatre was associated with better SRH among women and with better survival among men. After adjusting for several other cultural activities, the association with SRH among women disappeared, and the effect on survival among men was attenuated and could be due to chance. People who engaged in music, singing and theatre tended to be active in several leisure activities.

To control for predicted confounding and moderating effects of gender, four separate gender-specific hierarchical multiple regression models were conducted. The gender-specific analyses were based on former studies that reveal gender-dependent outcomes (Cuypers et al. 2012; Ekholm et al. 2016; Hansen et al. 2015; Trainor et al. 2010). After adjustment for other creative cultural activities, however, attenuation of the estimates was observed, and the gender-specific significant results disappeared for both outcomes (model 4). This may indicate that effects of different cultural activities may be difficult to disentangle.

The gender differences in the association between participation in music, singing and theatre and all-cause mortality is surprising. Although not many cohorts conducted gender-specific analysis, those, which have done so, suggest that both men and women will benefit from reduced risk of mortality by participating in music or singing. For example, Bygren et al. found that singing in a choir or making music was a significant determinant of survival for Swedish men and women (Bygren 1996). And in the Finnish working population, there were significant associations between leisure participation and mortality in the unadjusted and fully adjusted models; leisure activities reduced the risk of mortality for men and women, with the exception of the healthy women at the beginning of the follow-up (Hyyppä et al. 2006). However, while leisure activities included listening to recorded music drama, singing and among many other activities, there was no separate analysis by type of leisure, and all activities were lumped under an overall index of leisure, suggesting that all leisure activities contributed towards better survival. In our analysis we were able to show that even after adjustment for recreational activity outdoors, sports participation, club membership, parish work and dance activities, these particular activities had added benefits for men survival although the 95% confidence interval suggests that this can still be due to chance. Additionally, it should be noted that the association found here in men and not in women could be due to confounding by lifestyle factors, particularly alcohol, smoking, dietary factors, sleeping time or social networking. These factors are associated with mortality and more likely be differently distributed by gender, such as women tend to socialize more than men and smoke less than men (Steinar Krokstad et al. 2017). Hence, not accounting for these factors may bias our estimates. Although some data on lifestyle factors exist in the questionnaire from the HUNT3 survey, the item non-response rate was high for some of the questions on health-related behaviours. Including only participants with no single missing data would have reduced our sample substantially, which is one of the weaknesses of our report. Further analyses of possible confounders remain, but previous published studies show that these lifestyle factors have not influenced significantly (Bygren 1996; Väänänen et al. 2009) and therefore may not be of importance. Additionally, if cultural participation negatively correlates with health-related lifestyle factors, one would assume that the social relationships that accompany participation would be of importance. For example, lack of participation may negatively affect sleeping or other health-related behaviour because of something meaningful in life is missing. Opposite participation may buffer the effect of negative health-related lifestyle factors. However given that not many studies investigated gender differences, this was an interesting question on its own. Cuypers et al. also revealed that men participating in music, singing and theatre were significantly more likely to have lower depression scores. This may be the mechanism behind the findings here as depression and mortality are strongly associated (Cuijpers and Smit 2002).

Regarding SRH, the results of the present study are partly in accordance with another study conducted by Nummela et al. They found that women attending singing in a choir, doing art painting or playing music at least once a month had a better SRH than their less frequently participating counterparts. And women reported slightly better SRH compared to men (Nummela et al. 2008). However, the study was cross-sectional and did not involve younger men and women precluding full comparison with our findings. Other studies have found that those who are participating in cultural activities seem to be more satisfied with their lives (Cuypers et al. 2012; Wilkinson et al. 2007). Although satisfaction with life is not identical to SRH, many measures of life satisfaction, such as subjective well-being index, include the subdomain health, which found to have consistently unique and significant contribution to satisfaction with life as a whole among Australian population (Cummins et al. 2003).

In a Finnish study, the investigators also examined the association between leisure participation index and SRH, and contrary to our finding, the association was insignificant for women (Hyyppä et al. 2006). Results from the survey in Denmark conducted by Ekholm et al. did not reveal any difference between women and men in the association with participation in music and SRH (Ekholm et al. 2016). In general, few population studies are performed. In addition, one challenge is that the activities that are investigated are so different. Therefore it is difficult to directly compare between studies.

In a study conducted by Schou et al., the relationship between SRH and mortality was studied in a Norwegian population, The HUNT1 Survey (1984–1986). At group level, they found that those who did consider their health as poor had a significantly higher risk for mortality than those who consider their health as good. Gender-specific analysis did not prove any differences among men and women (Schou et al. 2006). Gender-dependent associations from our results are in agreement with the findings that Cuypers et al. found in the study from the same material (Cuypers et al. 2012). A statement in a Norwegian book on gender differences in a public health perspective states: "Women are sicker, but men die quicker" (Lorber and Moore 2002; Mæland and Haugland 2007). These gender differences in men's lower life expectancy and women's greater morbidity which impact on their SRH seem to have both biological and sociocultural connections which impact biological processes (Krieger 2003; Rieker and Bird 2005). Further research is needed to explore these mechanisms and underlying causes with respect to gender differences in the outcomes that are associated with similar exposures to music, singing and theatre.

What dose of culture engagement is necessary to achieve health effects that may be an important issue in this context (Knudtsen et al. 2005)? Our results suggest that people who never engage are at the highest risk for health problems. Moreover, the dose of participation needs not to be necessarily very high. The importance of cultural participation is probably being overlooked in the public health perspective and in the context of treatment (Knudtsen et al. 2005). Many of those who suffer from mental and physical illness lack both experiences and networks (Knudtsen et al. 2005). Cultural activities may directly influence upon health through palliative coping or substituting health-compromising behaviours. Cultural engagement may also facilitate the development of social networks, which can improve health via social support and the dissemination of social health norms (Thiel 2015). A study examined the role of cultural leisure time activities in the work stress recovery of hospital employees (Tuisku et al. 2016). Both receptive and creative activities and frequency of cultural activities were analysed. Employees who reported both types of activities on a weekly basis had the highest relaxation mastery and control experiences during off-job time. Particular creative activities played an important role in certain aspects of recovery (Tuisku et al. 2016). Theorell et al. published a Swedish study that among 8000 twins found a significant association between musical management and emotional competence (Theorell et al. 2014). Music foster feelings of social connection, and music can be used to regulate or initiate feelings like joy, sadness, peacefulness, fear or tranquillity (Chanda and Levitin 2013).

In the present study, participation in music, singing and theatre is an active creative attendance, defined as performing arts (Davies et al. 2012). The definition of art varies considerably by the art form and type of activities and is often inconsistent (Cuypers et al. 2011; Davies et al. 2012), and this makes it difficult to compare results between studies. In addition, differences in social structure and health outcomes between different continents, and to an extent within Scandinavia, challenge any comparison between the results of population-based studies. There may be many pathways from participating in music, singing and theatre to health. In agreement with Davies, further research is needed to quantify the arts-social and the arts-physical health relationship (Davies et al. 2016a). In agreement with previous research, more research is needed with the aim to investigate the association in general population and preferable prospective studies with multiple measurement over time (Bygren 1996; Cuypers et al. 2012; Cuypers et al. 2011; Ekholm et al. 2016; Nenonen et al. 2014; Wilkinson et al. 2007). Research conducted by Johansson et al. support this (Johansson et al. 2001). They performed a longitudinal Swedish interview study and investigated how changes in the habit of attending cultural events in a community might predict SHR. The findings were that those who became less culturally active between the first and second occasion, or were culturally inactive on both occasions, had a higher risk of impaired perceived health compared to those who were culturally active on both occasions. And the results could be in agreement with a casual influence of stimulation and suggest that cultural stimulation is a 'perishable commodity' (Johansson et al. 2001). The study of working subjects by Theorell et al. (2013) points in the same direction. That study was performed on actively working subjects who were not on sick leave, and the prospective 2-year results showed, after adjustment for confounders, that psychological health developed in a better way among working people who had cultural activities organized at work. It is likely that subjects on sick leave who have poor health could either participate more or less in cultural activities than others because of their illness. This may cause serious interpretation difficulties in this kind of research which need to be explored thoroughly (Theorell et al. 2013). New measurements on cultural participating will be completed with the HUNT4 study in 2017–2018, and this will give rise to the opportunities for accomplishing prospective longitudinal studies.

Strengths, Limitations and Methodological Issues

The strengths of our study include the population-based prospective design, the large sample size with good response rate (54%) and the availability of several relevant covariates. The data are collected in a health survey covering many aspects of health, ruling out reporting bias for the associations studied. Data from the third HUNT survey do have detailed quantification of arts engagement, both creative and receptive cultural activities (Holmen et al. 2016). The HUNT3 population is ethnically homogenous and has been considered fairly representative for the Norwegian population. The social inequalities in the population might be slightly smaller than for Norway in general, since Nord-Trøndelag lacks large cities (Krokstad et al. 2013; Krokstad and Westin 2002). In general, more women (58.7%) than men participated in the study. In the age group 60–69 with the highest participation, the rate was 74.5% for women and 67.7% for men (Krokstad et al. 2013). Participation rates rate in HUNT were lower among the younger age groups (20–39 years) than the older population, with the exception of those aged 80 years or above (Krokstad et al. 2013).

We excluded 7731 (18.8%) participants with incomplete answers on one or more variables. Out of these 4501 had incomplete response on the main exposure variable, but we have no way to estimate if this loss was at random on the main exposure measures (cultural activity) or we had some systematic loss.

A non-participant study from the HUNT population has been performed, and these analyses indicate that those who decline to participate had higher mortality than those who participated, and lower SES compared to participants, and higher prevalence of several chronic diseases. In addition, non-participants of HUNT surveys displayed higher mortality years after the survey was conducted (Langhammer et al. 2012).

A strength of the study is in the validity of the endpoint; all-cause mortality is considered to be very accurate because of the high quality of national registries in Norway (Bakken et al. 2015). The follow-up time, 7 years, may be considered as sufficiently long. Based on the findings in the focus group interviews, from the evaluation of the exposure variable (Holmen et al. 2016), there is reason to assume that those who responded to the cultural participation issues managed to distinguish between active and passive participation. The questions allow for different interpretations at the individual level, but no indications were found on systematic misinterpretations (Holmen et al. 2016). The exposure variable contains information about music, singing and theatre activities that cannot be separated from each other. And we cannot distinguish between the creative activity music, singing and theatre in itself or the social network and the relationship with others that is the cause of association (Aartsen et al. 2017). Socially shared and solitary cultural activities both may potentially extend lifespan through various pathways (Bygren 1996; Väänänen et al. 2009). Evidence from cross-sectional studies point out associations between participation and health outcomes but are not strong to suggest causality because of the lack of temporal association.

Longstanding illness, social network and SRH were measured with only one question. For SRH this is proved to be sufficient (Bowling 2005). SRH is a measure of people's perceptions of their subjective health (Huisman and Deeg 2010; Jylhä 2009). Cultural participation is associated with socioeconomic status (SES) and has been emphasized as an important topic to be aware of (Cuypers et al. 2012). A report on health and arts showed a clear association between SES and the likelihood of attendance at arts (Windsor 2005). The opportunities to be healthy and remain healthy are unequally distributed across SES groups (Aartsen et al. 2017). SES is strongly associated with risk of premature mortality and is associated with health on all levels of the hierarchy (Adler and Ostrove 1999; Adler et al. 1993). Health behaviours, the degree of control and emotional and instrumental support vary with different SES (Adler et al. 1993). We may not adjust properly for SES and social network, as we were only able to include education as a measure of SES. Had we included other variables it could have led, on the other hand, to over adjustment. We could have tested the estimates for other social factors, as social support is seen as one of the social determinants of health in the general population (Wilkinson and Marmot 2003). But there is lack of evidence in which direction the covariate affects both the exposure (attending music, singing and theatre) and the outcome. In addition, we adjusted for all the other creative activities that were measured in the questionnaire at baseline. The association disappeared when we adjusted for participation in the other health-enhancing physical activities such as sports or outdoor recreational activities. It could be that the effect on health is cumulative and related to the amount of activities participants engage in, rather than, a particular type.

Conclusion

Our data indicate that participating in music, singing and theatre seem to affect subjective SRH and all-cause mortality, but there are gender differences. Women who regularly participate in music, singing and theatre had an increased SRH, and men had a reduced all-cause mortality. Yet, it appeared that other leisure activities might have similar or greater effects on both outcomes, once all the other creative activities were taken into account.

Main Message

Cultural activities like playing an instrument, singing or participating in theatre can increase SRH in women and protect against all-cause mortality in men. Stimulating such activities may have positive health effects in the population.

Acknowledgement The Nord-Trøndelag Health Study (The HUNT Study) is a collaboration between HUNT Research Centre, (Faculty of Medicine and Health Sciences, NTNU, Norwegian University of Science and Technology), Nord-Trøndelag County Council, Central Norway Health Authority and the Norwegian Institute of Public Health.

References

Aartsen, M., Veenstra, M., & Hansen, T. (2017). Social pathways to health: On the mediating role of the social network in the relation between socio-economic position and health. *SSM – Population Health, 3*, 419–426. https://doi.org/10.1016/j.ssmph.2017.05.006.

Adler, N., & Ostrove, J. (1999). Socioeconomic status and health: What we know and what we don't. *Annals of the New York Academy of Science, 896*, 3–15.

Adler, N. E., Boyce, W., Chesney, M. A., Folkman, S., & Syme, S. (1993). Socioeconomic inequalities in health: No easy solution. *JAMA, 269*(24), 3140–3145. https://doi.org/10.1001/jama.1993.03500240084031.

Bakken, I. J., Ellingsen, C. L., Pedersen, A. G., Leistad, L., Kinge, J. M., Ebbing, M., & Vollset, S. E. (2015). Comparison of data from the cause of death registry and the Norwegian patient register. *Tidskr Nor Lægeforen, 135*(21), 19439. https://doi.org/10.4045/tidsskr.14.0847.

Bowling, A. (2005). Just one question: If one question works, why ask several? *Journal of Epidemiology & Community Health, 59*(5), 342–345.

Bygren, L. O., Konlaan, B., & Johansson, S. E. (1996). Attendance at cultural events, reading books or periodicals, and making music or singing in a choir as determinants for survival: Swedish interview survey of living conditions. *BMJ, 313*, 1577.

Chanda, M. L., & Levitin, D. J. (2013). The neurochemistry of music. *Trends in Cognitive Sciences, 17*(4), 179–193. https://doi.org/10.1016/j.tics.2013.02.007.

Cuijpers, P., & Smit, F. (2002). Excess mortality in depression: A meta-analysis of community studies. *Journal of Affective Disorders, 72*(3), 227–236. https://doi.org/10.1016/S0165-0327(01)00413-X.

Cummins, R. A., Eckersley, R., Pallant, J., Van Vugt, J., & Misajon, R. (2003). Developing a national index of subjective wellbeing: The Australian Unity Wellbeing Index. *Social Indicators Research, 64*(2), 159–190. https://doi.org/10.1023/A:1024704320683.

Cuypers, K. F., Knudtsen, M. S., Sandgren, M., Krokstad, S., Wikström, B. M., & Theorell, T. (2011). Cultural activities and public health: Research in Norway and Sweden. An overview. *Arts & Health, 3*(1), 6–26. https://doi.org/10.1080/17533015.2010.481288.

Cuypers, K., Krokstad, S., Lingaas Holmen, T., Skjei Knudtsen, M., Bygren, L. O., & Holmen, J. (2012). Patterns of receptive and creative cultural activities and their association with perceived health, anxiety, depression and satisfaction with life among adults: The HUNT study, Norway. *Journal of Epidemiology and Community Health, 66*(8), 698. https://doi.org/10.1136/jech.2010.113571.

Davies, C., Rosenberg, M., Knuiman, M., Ferguson, R., Pikora, T., & Slatter, N. (2012). Defining arts engagement for population-based health research: Art forms, activities and level of engagement. *Arts Health, 4*, 203–216. https://doi.org/10.1080/17533015.2012.656201.

Davies, C. R., Knuiman, M., Wright, P., & Rosenberg, M. (2014). The art of being healthy: A qualitative study to develop a thematic framework for understanding the relationship between health and the arts. *BMJ Open, 4*(4), e004790. https://doi.org/10.1136/bmjopen-2014-004790.

Davies, C., Knuiman, M., & Rosenberg, M. (2016a). The art of being mentally healthy: A study to quantify the relationship between recreational arts engagement and mental well-being in the general population. *BMC Public Health, 16*(1), 15. https://doi.org/10.1186/s12889-015-2672-7.

Davies, C., Pescud, M., Anwar-McHenry, J., & Wright, P. (2016b). Arts, public health and the National Arts and health framework: A lexicon for health professionals. *Australian and New Zealand Journal of Public Health, 40*(4), 304–306. https://doi.org/10.1111/1753-6405.12545.

Daykin, N., Orme, J., Evans, D., Salmon, D., McEachran, M., & Brain, S. (2008). The impact of participation in performing arts on adolescent health and behaviour. *Journal of Health Psychology, 13*(2), 251–264. https://doi.org/10.1177/1359105307086699.

Diba, D., & D'Oliveira, A. F. (2015). Community theater as social support for youth: Agents in the promotion of health. *Ciência & Saúde Coletiva, 20*(5), 1353. https://doi.org/10.1590/1413-81232015205.01542014.

Ekholm, O., Juel, K., & Bonde, L. O. (2016). Associations between daily musicking and health: Results from a nationwide survey in Denmark. *Scandinavian Journal of Public Health, 44*(7), 726–732. https://doi.org/10.1177/1403494816664252.

Gold, C., Solli, H. P., Krüger, V., & Lie, S. A. (2009). Dose-response relationship in music therapy for people with serious mental disorders: Systematic review and meta-analysis. *Clinical Psychology Review, 29*(3), 193. https://doi.org/10.1016/j.cpr.2009.01.001.

Gold, C., Assmus, J., Hjørnevik, K., Qvale, L. G., Brown, A. F. K., Waage, L., … Stige, B. (2014). Music therapy for prisoners: Pilot randomised controlled trial and implications for evaluating psychosocial interventions. International Journal of Offender Therapy and Comparative Criminology, 58(12), 1520–1539. https://doi.org/10.1177/0306624X13498693

Grape, C., Sandgren, M., Hansson, L.-O., Ericson, M., & Theorell, T. (2002). Does singing promote well-being?: An empirical study of professional and amateur singers during a singing lesson. *Integrative Physiological & Behavioral Science, 38*(1), 65–74. https://doi.org/10.1007/BF02734261.

Hansen, E., Sund, E., Knudtsen, M. S., Krokstad, S., & Holmen, T. L. (2015). Cultural activity participation and associations with self-perceived health, life-satisfaction and mental health:

The young HUNT study, Norway. *BMC Public Health, 15*(1), 1–8. https://doi.org/10.1186/s12889-015-1873-4.

Holmen, J., Midthjell, K., Kruger, O., Langhammer, A., Holmen, T., & Bratberg, G. H. (2003). The Nord-Trøndelag health study 1995–97 (HUNT2): Objectives, contents, methods and participation. *Norwegian Journal of Epidemiology, 13*, 19–32.

Holmen, J., Nguyen, C., Haapnes, O., Rangul, V., & Espnes, G. A. (2016). Kultur og helse i HUNT – En metodeevaluering. *Norwegian Journal of Epidemiology, 26*(1–2), 139–144.

Holt-Lunstad, J., Smith, T. B., Layton, J. B., & Brayne, C. (2010). Social relationships and mortality risk: A meta-analytic review (social relationships and mortality). *PLoS Medicine, 7*(7), e1000316. https://doi.org/10.1371/journal.pmed.1000316.

Huisman, M., & Deeg, D. J. H. (2010). A commentary on Marja Jylhä's "what is self-rated health and why does it predict mortality? Towards a unified conceptual model"(69:3, 2009, 307–316). *Social Science & Medicine, 70*(5), 652–654. https://doi.org/10.1016/j.socscimed.2009.11.003.

HUNT Research Centre. (2017a). HUNT databank, NTNU. Norwegian national registry from (HUNT). https://hunt-db.medisin.ntnu.no/hunt-db/#/instrument/229.

HUNT Research Centre. (2017b). Research. NTNU. Merging HUNT data with other Norwegian registry data. from NTNU http://www.ntnu.edu/hunt/merging-registries.

Hyyppä, M. T., Mäki, J., Impivaara, O., & Aromaa, A. (2006). Leisure participation predicts survival: A population-based study in Finland. *Health Promotion International, 21*(1), 5. https://doi.org/10.1093/heapro/dai027.

Johansson, S. E., Konlaan, B. B., & Bygren, L. O. (2001). Sustaining habits of attending cultural events and maintenance of health: A longitudinal study. *Health Promotion International, 16*(3), 229–234.

Jylhä, M. (2009). What is self-rated health and why does it predict mortality? Towards a unified conceptual model. *Social Science & Medicine, 69*(3), 307–316. https://doi.org/10.1016/j.socscimed.2009.05.013.

Knudtsen, M. S., Holmen, J., & Håpnes, O. (2005). Cultural approaches to treatment and public health work. *Tidsskrift for den Norske lægeforening : tidsskrift for praktisk medicin, ny række, 125*(24), 3434.

Konlaan, B., Bygren, L., & Johansson, S. E. (2000). Visiting the cinema, concerts, museums or art exhibitions as determinants of survival: A Swedish fourteen-year cohort follow-up. *Scandinavian Journal of Public Health, 28*, 128–178.

Krieger, N. (2003). Genders, sexes, and health: What are the connections – and why does it matter? *International Journal of Epidemiology, 32*(4), 652–657.

Krokstad, S., & Westin, S. (2002). Health inequalities by socioeconomic status among men in the Nord-Trondelag health study, Norway. *Scandinavian Journal of Public Health, 30*(2), 113–124. https://doi.org/10.1080/14034940210133753.

Krokstad, S., Langhammer, A., Hveem, K., Holmen, T., Midthjell, K., Stene, T., … Holmen, J. (2012). Cohort Profile: The HUNT study, Norway. *International Journal of Epidemiology, 42*. doi: https://doi.org/10.1093/ije/dys095.

Krokstad, S., Langhammer, A., Hveem, K., Holmen, T. L., Midthjell, K., Stene, T. R., … Holmen, J. (2013). Cohort Profile: The HUNT Study, Norway. *International Journal of Epidemiology, 42*(4), 968–977. doi: https://doi.org/10.1093/ije/dys095

Krokstad, S., Ding, D., Grunseit, A. C., Sund, E. R., Holmen, T. L., Rangul, V., & Bauman, A. (2017). Multiple lifestyle behaviours and mortality, findings from a large population-based Norwegian cohort study – the HUNT Study.(Nord-Trøndelag Health Study)(Report). *BMC Public Health, 17*(1), 58. https://doi.org/10.1186/s12889-016-3993-x.

Langhammer, A., Krokstad, S., Romundstad, P., Heggland, J., & Holmen, J. (2012). The HUNT study: Participation is associated with survival and depends on socioeconomic status, diseases and symptoms. *BMC Medical Research Methodology, 12*(1), 1–14. https://doi.org/10.1186/1471-2288-12-143.

Li, S., Stampfer, M. J., Williams, D. R., & Vanderweele, T. J. (2016). Association of religious service attendance with mortality among women. *JAMA Internal Medicine, 176*(6), 777. https://doi.org/10.1001/jamainternmed.2016.1615.

Llovet, A. K. (2017). Meta-analysis of specific music therapy measures and their implications for the health care system. *Health Care Management (Frederick), 36*(1), 60–68. https://doi.org/10.1097/hcm.0000000000000135.

Lorber, J., & Moore, L. J. (2002). Women get sicker but men die quicker: Social epidemiology – in – gender and the social construction of illness. In J. Lorber & L. J. Moore (Eds.), *Gender and the social construction of illness* (2nd ed., pp. 13–36). Lanham: AltaMira.

Mæland, G., & Haugland, S. (2007). Det syke kjønn? Sykelighet blant kvinner og menn i Norge. In B. Schei & L. S. r. Bakketeig (Eds.), *Kvinner lider – Menn dør. Folkehelse i et kjønnsperspektiv.* (Vol. 1, pp. 69–87): Gyldendal Akademisk.

McEwen, B. S. (2006). Protective and damaging effects of stress mediators: Central role of the brain. *Dialogues in Clinical Neuroscience, 8*(4), 367–381.

McEwen, B. (2012). Brain on Stress: How the social environment gets under the skin. *Proceedings of the National Academy of Sciences, 109*(Suppl2), 17180–17185.

Merom, D., Ding, D., & Stamatakis, E. (2016). Dancing participation and cardiovascular disease mortality. A pooled analysis of 11 population-based British cohorts. *American Journal of Preventive Medicine, 50*(6), 756–760.

Nenonen, T., Kaikkonen, R., Murto, J., & Luoma, M.-L. (2014). Cultural services and activities: The association with self-rated health and quality of life. *Arts & Health, 6*(3), 235–253. https://doi.org/10.1080/17533015.2014.897958.

Nummela, O., Sulander, T., Rahkonen, O., & Uutela, A. (2008). Associations of self-rated health with different forms of leisure activities among ageing people. *International Journal of Public Health, 53*(5), 227–235. https://doi.org/10.1007/s00038-008-6117-2.

Raglio, A., Filippi, S., Bellandi, D., & Stramba-Badiale, M. (2014). Global music approach to persons with dementia: Evidence and practice. *Clinical Interventions in Aging*, 1669–1676.

Rieker, P. P., & Bird, C. E. (2005). Rethinking gender differences in health: Why we need to integrate social and biological perspectives. *The Journals of Gerontology. Series B, Psychological Sciences and Social Sciences, 60.* Spec No.(2), 40–47.

Schou, M. B., Krokstad, S., & Westin, S. (2006). How is self-rated health associated with mortality? *Tidsskrift for den Norske Lægeforening, 126*(20), 2644–2647.

Theorell, T., Liljeholm-Johansson, Y., Bjork, H., & Ericson, M. (2007). Saliva testosterone and heart rate variability in the professional symphony orchestra after "public faintings" of an orchestra member. *Psychoneuroendocrinology, 32*(6), 660–668. https://doi.org/10.1016/j.psyneuen.2007.04.006.

Theorell, T., Osika, W., Leineweber, C., Magnusson Hanson, L., Bojner Horwitz, E., & Westerlund, H. (2013). Is cultural activity at work related to mental health in employees? *International Archives of Occupational and Environmental Health, 86*(3), 281–288. https://doi.org/10.1007/s00420-012-0762-8.

Theorell, T. P., Lennartsson, A.-K., Mosing, M. A., & Ullén, F. (2014). Musical activity and emotional competence – a twin study. *Frontiers in Psychology, 5*, 774. https://doi.org/10.3389/fpsyg.2014.00774.

Thiel, L. (2015). Leave the drama on the stage: The effect of cultural participation on health. *SOEPpapers on Multidisciplinary Panel Data Research.* Berlin. German Socio-Economic Panel Study (SOEP). https://www.diw.de/documents/publikationen/73/diw_01.c.510133.de/diw_sp0767.pdf

Trainor, S., Delfabbro, P., Anderson, S., & Winefield, A. (2010). Leisure activities and adolescent psychological well-being. *Journal of Adolescence, 33*(1), 173–186. https://doi.org/10.1016/j.adolescence.2009.03.013.

Tuisku, K., Virtanen, M., De Bloom, J., & Kinnunen, U. (2016). Cultural leisure activities, recovery and work engagement among hospital employees. *Industrial Health, 54*(3), 254–262. https://doi.org/10.2486/indhealth.2015-0124.

Väänänen, A., Murray, M., Koskinen, A., Vahtera, J., Kouvonen, A., & Kivimäki, M. (2009). Engagement in cultural activities and cause-specific mortality: Prospective cohort study. *Preventive Medicine, 49*(2), 142–147. https://doi.org/10.1016/j.ypmed.2009.06.026.

Verghese, J., Lipton, R. B., Katz, M. J., Hall, C. B., Derby, C. A., Kuslansky, G., … Buschke, H. (2003). Leisure activities and the risk of dementia in the elderly. *The New England Journal of Medicine, 348*(25), 2508.

Wilkinson, R., & Marmot, M. (2003). Social determinants of health: the solid facts. 2nd edition. In R. Wilkinson & M. Marmot (Eds.). Europe.

Wilkinson, A. V., Waters, A. J., Bygren, L. O., & Tarlov, A. R. (2007). Are variations in rates of attending cultural activities associated with population health in the United States? *BMC Public Health, 7*, 226–226. https://doi.org/10.1186/1471-2458-7-226.

Windsor, J. (2005). *Your health and the arts: A study of the association between arts engagement and health*. London: Arts Council England.

Chapter 4
Music Practice and Emotion Handling

Töres Theorell and Fredrik Ullén

Introduction

It is widely recognized that musical experiences may evoke different kinds of emotions and also amplify emotions in a given moment (Theorell 2014; Gabrielsson 2011; Juslin and Sloboda 2010; Ruud 2010). However, questions relating to possible relationships of repeated musical experiences to emotional skills have not been studied extensively. In this chapter, we shall explore such relationships and also the extent and nature of such relationships in relation to health and creativity. To put the questions in simpler language:

Is there any correlation between a life with music on one hand and ability to handle emotions on the other hand?

To interpret and differentiate, as well as to describe and communicate emotions, is of great importance for social interactions and for well-being and consequently also for health. In psychosomatic medicine, difficulties in such abilities are labelled *alexithymia*. The term alexithymia was introduced by Sifneos (1973, 1996) and has been further developed by several authors (Nemiah 1996). An alexithymic person has difficulties in the communication of feelings. This may lead to poor communication with others. It seems to other people that these individuals minimize emotional experience since they seem to focus their attention externally and not on their emotions. They also tend to have reduced capacity for fantasizing and symbolic thinking. Therefore, alexithymia is potentially of significance to creativity.

T. Theorell (✉)
Department of Neuroscience, Karolinska Institute, Stress Research Institute, Stockholm University, Stockholm, Sweden
e-mail: Tores.Theorell@ki.se

F. Ullén
Department of Neuroscience, Karolinska Institutet, Stockholm, Sweden
e-mail: Fredrik.Ullen@ki.se

L. O. Bonde, T. Theorell (eds.), *Music and Public Health*,
https://doi.org/10.1007/978-3-319-76240-1_4

Alexithymia is related to structural deviances in emotional systems of the brain (Borsci et al. 2009; Laricchiuta et al. 2014), as well as to deviating patterns of regional brain activity during the processing of emotional stimuli (Deng et al. 2013; Duan et al. 2010; Heinzel et al. 2010), mentalizing (Moriguchi et al. 2006) and imagery (Mantani et al. 2005). There is evidence for a genetic contribution (Jorgensen et al. 2007) to alexithymia, but environmental factors also play a role. Alexithymia is associated with several outcomes related to health, such as depression (Li et al. 2015), hypertension (Grabe et al. 2010; Jorgensen and Houston 1986), sympathetic overactivity (Fukunishi et al. 1999), somatic complaints and symptoms (De Gucht and Heiser 2003), as well as to lowered quality of life (Mattila et al. 2009, 2010). The most commonly used instrument for assessing alexithymia is the Toronto Alexithymia Scale, TAS-20, (Bagby et al. 1994). According to the assessment theory underlying this questionnaire, alexithymia is not a distinct clinical diagnosis but rather a dimensional trait. Alexithymia is more prevalent among men than among women and is negatively associated with years of education (Salminen et al. 1999).

Several studies have shown stability of alexithymia over time in both depressed patients receiving treatment (Luminet et al. 2001; Salminen et al. 1994) and the general population (Tolmunen et al. 2011). On the other hand, intervention studies have shown that difficulty to handle emotions may be reduced, such as in a study on depressed patients who received medication and/or psychotherapy (Honkalampi et al. 2000). There are indications that participating in artistic and cultural activities in general (in other words, not only music) may reduce alexithymia. In a randomized controlled intervention study (see also Chap. 12 by Horwitz), women with burnout symptoms were allocated to either participate in cultural activities such as dance, theatre, drawing and vocal improvisation during a 3-month period or to belong to a waiting-list control group receiving "usual care" in primary care settings (Grape Viding et al. 2015). Alexithymia scores decreased markedly and significantly in the intervention group during the intervention period and continued to do so during 3 months of follow-up, while the control group showed stable alexithymia scores during the intervention and follow-up. It is likely that practicing activities such as music, dance, theatre, visual arts and writing may amplify emotional experiences and facilitate translations of emotions and thereby reduce alexithymia. However, studies in this area are very limited (Bojner Horwitz et al. 2015; Theorell et al. 2014) and not specific to music.

Cultural Activity and Alexithymia

The aim of the first group of studies was to investigate whether there is an association between creative artistic activities, in particular music practice, and alexithymia. In the first study (Theorell et al. 2014), we used the participants' own estimation of the number of hours of music practice in different ages with the accumulated total life sum of hours as our main explanatory variable. The expression and perception of emotions are indeed central elements of music listening and performance (Juslin and Sloboda 2010). This relates to a more profound and complex question about the functional role of music for human beings and what role emotions may have in this (Madison 2011). According to some authors, music may facilitate the experience of emotions and thus be of importance for the development of the ability to identify and differentiate emotions (Dissanayake 2000; Laiho 2004; Baumgartner et al. 2006; Gabrielsson 2011). Following this reasoning, music experiences might lower the risk of alexithymia. If musical engagement could facilitate emotional development, music education may potentially serve as an intervention strategy to reduce alexithymia, particularly in adolescence.

The study sample was recruited from the Swedish Twin Registry (Theorell et al. 2014). The participants were in the ages 27–54. All analyses were adjusted for age. In the genetic analysis, there were 8110 subjects who had valid information about zygosity, alexithymia score and number of hours of music practice. When a twin population is used for studies in the same way as a population study, it is a problem that the two partners in a twin pair are more similar to one another than are members of the general population. In order to handle this problem, the so-called phenotypic analyses are made. For the present phenotypic analyses, to control for relatedness within the sample, we randomly selected as participants from complete pairs only one twin from each pair. In addition, all single twins (whose partner had not participated) were included.

Men had higher alexithymia scores than women, and subjects with a low educational level had higher alexithmia scores than others. Intelligence, as measured with the Wiener Matrizen Test, was not correlated with alexithymia, neither in men nor in women. Both number of music practice hours and musical achievement (scored from 1, no experience of playing music, to 7, nationally or internationally awarded professional; see Fig. 4.1 below) were significantly negatively correlated with alexithymia – the more achievement in music and the higher number of music practice hours, the less likelihood of being alexithymic. The amount of explained variance was in the order of 1%.

An interesting finding was that among those who reported ensemble playing, there was an additional statistically significant effect on the alexithymia score. Those who reported a high number of hours of music practice and who had practiced ensemble had a particularly low risk of having a low alexithymia score. This is displayed in Fig. 4.1 for men and in Fig. 4.2 for women. The participants were divided into tertiles (horizontal axis) with regard to lifetime number of hours of music practice. The vertical axis displays the average alexithymia score. Subjects without ensemble experience correspond to the dotted line. The difference between those in the upper and those in the lower tertile was in the order of three units on the alexithymia scale – both among men and among women. If the experience of ensemble playing is added to this, the difference increased to 4.5 both among men and among women. Since the standard deviation is ten units, this corresponds to nearly half of a standard deviation. This difference both among men and among women, respectively, was approximately of the same magnitude as the difference between men and women.

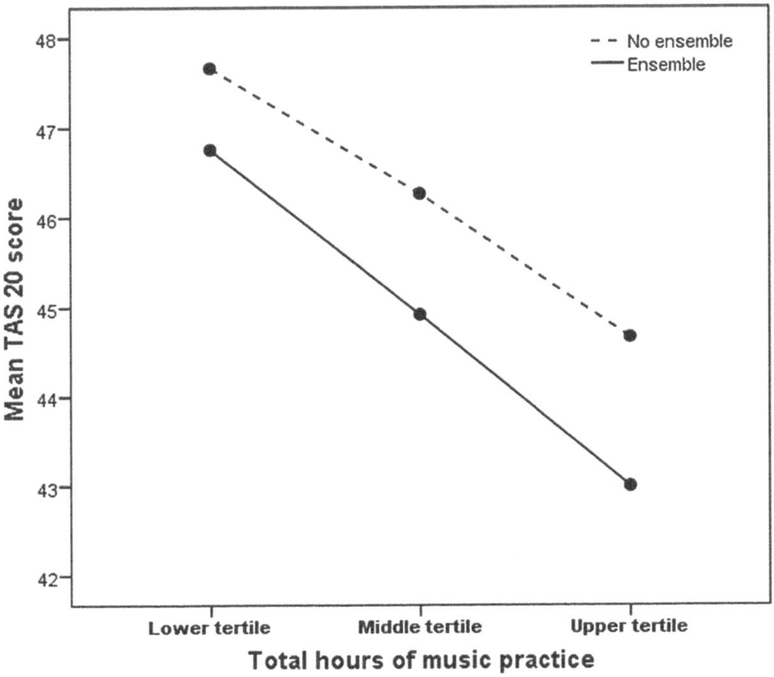

Fig. 4.1 Age-adjusted marginal Toronto Alexithymia Scale (TAS-20) means for ensemble (yes/no) and music practice tertiles among subjects (men) who have practiced playing an instrument. Music practice tertiles: lower, range 40–800, median 360; middle, range 840–3160, median 1600; upper, range 3200–18,400, median 5600. (From Theorell et al. (2014))

The genetic analysis showed that the association between the number of hours of music practice and alexithymia was indirect, an expression of *pleiotropy*. Both motivation to spend many hours practicing music and alexithymia are partly genetic (in the case of alexithymia in the order of 30%). When these associations were combined, it became obvious that there was no direct causal association between the number of hours of music practice and alexithymia.

In summary, this first study showed that a high level of musical activities throughout life, particularly if there has been ensemble playing or singing, was associated with a lowered level of alexithymia. In other words, musicians are, in general and particularly if they have had extensive experience of ensemble playing or singing, better at handling emotions than others. This association only explains a small amount of variation. However on the total population level, it may still be important. We also found that on a population level, the association between music practice and emotional skill was strongly influenced by genetic factors. In other words, we should not have unrealistic expectations on the general effect of music education on emotion handling in our society.

In the next study – also based upon the study of music experiences in the Swedish Twin Registry – we examined the statistical correlation between achievement in

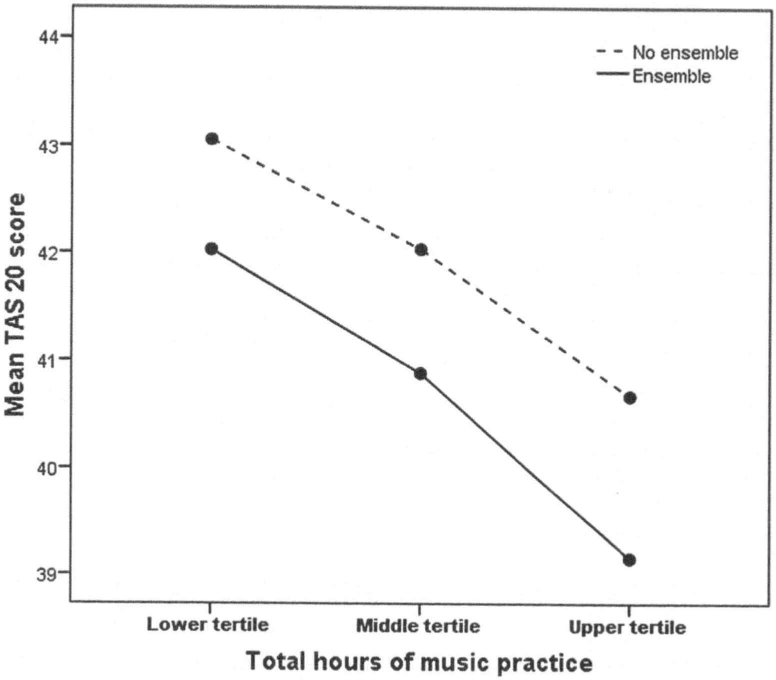

Fig. 4.2 Age-adjusted marginal TAS-20 means for ensemble (yes/no) and music practice tertiles among subjects (women) who have practiced playing an instrument. Music practice tertiles: lower, range 40–800, median 440; middle, range 840–2360, median 1360; upper, range 2400–20,800, median 5040. (From Theorell et al. (2014))

cultural activities in general and alexithymia. We examined whether different kinds of cultural achievement (i.e. in writing, music, visual arts, theatre and dance) could contribute statistically independently of one another to the alexithymia score – with the assumption that cultural achievements are likely to protect against alexithymia. We also tested whether these cultural achievements might add statistically to one another in predicting alexithymia score.

We used a Swedish version of the Creative Achievement Questionnaire in which participants self-estimate their lifelong achievement in creative activities on a scale ranging from no activity at all to professional activity reaching national or international recognition (Carson et al. 2005). All analyses were adjusted for age and educational level. In these phenotypic analyses, there were 2279 men and 3152 women.

The results showed both for men and women that high achievement in writing and music contributes statistically independently of one another and of other artistic achievements to a low alexithymia score. In men, achievement in visual arts and, in women, achievement in theatre also contributed to a low alexithymia score. In addition, there was evidence for an added effect of several concomitant achievements (Lennartsson et al. 2017).

A similar study of the relationship between artistic achievement in dancing and alexithymia was also performed in the same twin cohort. This is reported more in detail in another chapter in this book (see Chap. 12 by Bojner Horwitz). The results showed that both in men and in women, artistic achievement in dance was not significantly associated with alexithymia score, but one of the three subcomponents was, namely, externally oriented thinking. In other words, the more achievement in dance the more internally oriented thinking. The interpretation is that artistic achievement in dance is associated with high awareness of internal processes, and this may also result in a good ability to communicate such processes.

The phenotypic analyses of artistic achievement in general have accordingly shown that there are statistically significant relationships between artistic achievement in general and the ability to handle emotions and that concomitant achievements in several forms of art add to one another from this point of view. Writing should be mentioned as particularly important from this point of view. Dancing seems to be related to good awareness of internal processes. As with achievement in music, genetics probably contribute substantially to these relationships.

Interviews About Piano Playing Adjusting for Genetic Factors

Since genetic factors play an important role in the relationships between cultural activities and ability to handle emotions, an important step for us was to analyse a situation in which genetic factors are accounted for, namely, in interviews with monozygotic twins that in their lives became discordant with regard to music achievement and pair with one partner who has been playing piano throughout life, while the other partner became a nonplayer (Eriksson et al. 2016). Discordance throughout life was operationally defined as a difference in the number of hours of practice of at least 1000 h. Ten such identical twin pairs were recruited from the previously mentioned twin cohort. This difference means that all of the playing twins had substantial and ongoing experience of piano playing, while the nonplaying cotwins had played some piano as a child but then stopped entirely in adulthood. A semi-structured interview was performed with specified themes in defined segments of the interview. The first section dealt with perceived reasons for the discordance in music engagement. The second section was about specific within-pair differences in musical childhood environment, the third one about strong music memories and the fourth about the meaning of music in life and for well-being. There was also a subsequent fifth music-neutral segment – linguistic interest. These interviews gave no indication that the differences in musical engagement were caused by systematic environmental differences that were consistent across twin pairs. On the contrary, the twins provided a wide range of different explanations for the discordance.

These included differences in access to a piano, either while growing up or in adult life, differences in attitudes towards the musical genres performed, differences in the perceived need for a creative hobby, differences in social activities and musical interests among friends, different feelings about the common music teacher and different attitudes towards music as an expression of religious faith. One playing twin mentioned that the musical engagement of one of her peers may have been an influence on her own playing, although no relevant differences in peer interests were recalled by the other nine pairs. In another pair, the musically interested mother may have been a more important role model for the playing twin than for the nonplaying twin. In a third pair, the playing twin was reported to play more in public during childhood. Strong musical memories included unique events such as the negative memory of a failed public performance in one nonplaying twin.

Interestingly, however, the playing twins had a significantly higher mean in the music subscale of proneness to experience flow than their cotwins. This illustrates that their music practice had been associated with musical flow experiences. The statements regarding the meaning of music in life were more precise and more related to flow experiences and creativity in the playing twin than in the nonplaying. The latter partner was more conventional in his/her statements regarding the meaning of music.

ECG was recorded continuously during the twin interviews. The rationale behind this was that we wanted to know whether there would be "hot spots" characterizing emotional reactions either in the playing or in the nonplaying twin while talking about music experiences. ECG provides an interesting possibility to examine this since heart rate variability (HRV) is very sensitive to changes in the balance between the activation of the sympathetic (arousal) and the parasympathetic (slowing down) nervous system (Porges and Byrne 2012). An interesting difference arose in the start of the interview: there was a statistically significant difference in the time spent discussing the first question; the nonplaying twin spent 2.2 min and the playing twin only 1.4 min discussing why he/she believed the difference arose. Maybe this could be regarded as a "hot spot" which engaged the nonplaying twin. In the other segments of the interview, there was no such difference. In addition (see Fig. 4.3), there was a statistically significant increase in the "low-frequency (LF) power" part of heart rate variability in the nonplaying twin during this initial part of the interview. LF reflects a mixture of parasympathetic and sympathetic activation. In this particular case (since genetic factors are adjusted for and since our data showed that there was no difference at all in the parasympathetic part, high-frequency (HF) power of the HRV), an increase in LF reflects an increase in sympathetic activity. It is a very sensitive measure. There was no intrapair difference with regard to low-frequency power during the baseline period before the music interview and no such difference during the subsequent discussion about language interests. The LF differences developed significantly ($p = 0.02$) differently across time in the twin pairs during the music discussions. A possible interpretation of this could be that the nonplaying twin showed a higher level of physiological arousal and was more engaged by the question about dissimilarity in music experiences during childhood than the playing partner. This could be an illustration of possible lifelong consequences of exposure to music practice.

Ten playing "piano-discordant" monozygotic twin pairs. Measure in playing ("playing") minus corresponding measure in nonplaying ("nonplaying") twin. Logarithmically (2-log) transformed measures of low-frequency power in heart rate

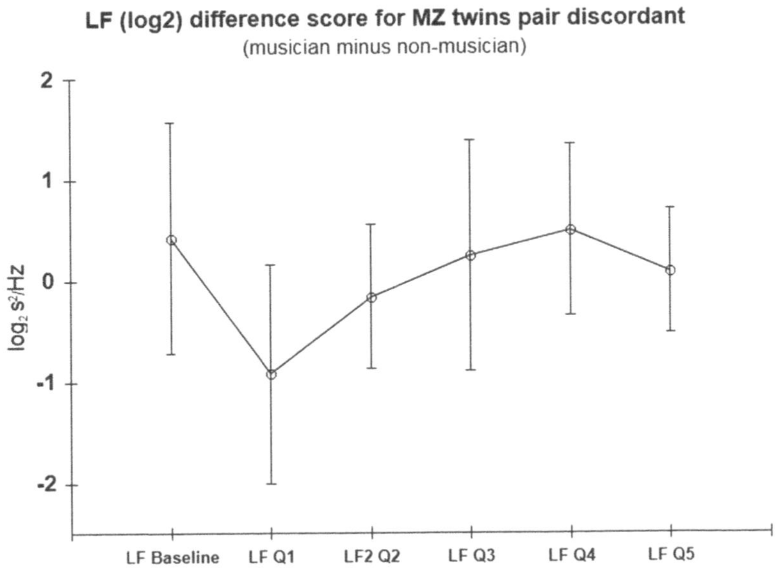

Fig. 4.3 Talking about music experiences in life in "piano discordant" identical twin pairs – sympathetic arousal differences. The *y*-axis corresponds to the mean difference (with standard deviation) between the 2log(LF power) in the playing and nonplaying twin (playing minus nonplaying) for each segment of the interview. The *x*-axis shows the different parts of the interview. Minus 1 corresponds to an excess of one unit in the nonplaying over the playing twin

variability during resting period before interview start (baseline) and subsequent interview segments, Q1, perceived reasons for the discordance in music engagement; Q2, specific within-pair differences in musical childhood environment; Q3, strong music memories; Q4, meaning of music in life and for the well-being; Q5, linguistic interest

Childhood Factors Predicting Continued Playing

A question arising from our findings on the observed relationships between music practice and artistic achievement in music on one hand and alexithymia on the other hand is whether childhood circumstances could predict the maintenance of music practice up to adult life (Theorell et al. 2015). Among those (aged 27–54 at the time of the web survey) who reported extra music education (added to the regular school curriculum), the following factors were significantly and strongly (odds ratio 1.5 or more corresponding to at least 50% increase in likelihood) predictive of continued playing until adulthood in multivariate analysis (Table 4.1):

Apart from male gender, all the factors that have a predictive value are environmental conditions that in principle could be stimulated if we want our children to continue

Table 4.1 Factors independently predicting continued music practice until adult years (age 27–54)

Boys more likely than girls to continue as adults
More than five persons in childhood family/friends practicing music
Classical music
Pop music
Special music class
Church context
Private education
Own choice of instrument
Lessons more than once a week
Early start

Swedish twin registry, $n = 3820$. Phenotypic multivariate analysis with mutual adjustment for explanatory variables including age

with music practice as adults, although one should keep in mind that similarities between parents and children in musical interests could depend on both genetic and environmental factors and, furthermore, that the response to environmental support for music is likely to be modulated by gene-environment interactions. Although in the total population there is a strong selection of children with basic musical talent, the environmental factors could play an important role. Gene-environment interaction has been suggested to be crucial in the development of musical expertise (Ullén et al. 2016). To say it in a simpler way, musical talent and ability to handle emotions associated with such talent probably have to be cultivated by the environment, and in the predictor study, we could identify several such factors.

Creative Jobs: Are Musical People More Likely to Work in Creative Jobs?

A question related to the discussion above is whether musicality and ability to handle emotions are associated with likelihood of working in creative occupations (Theorell et al. 2017). In order to illuminate this question, twin participants were subjected to an online survey, including computer-administered tests of musical aptitude (auditory musical discrimination), an intelligence test and the test for alexithymia (inability to handle emotions) described above. In this study, it was particularly relevant to use assessments of objective musicality tests. The musicality tests used in the study were the Swedish Musical Discrimination Test (Ullén et al. 2014), which includes three subscales for pitch (ability to differentiate high and low notes), rhythm (ability to repeat different rhythms) and melody (ability to recognize melodies). Participants also provided their educational level and a free-text job description, which was coded according to the O*NET system (http://www.onetcenter.org/overview.html). This has been based upon epidemiological interrogation about how people working in different occupations assess a number of aspects of their job. One such aspect is creativity or amount of creative thinking at work. The level of creativity for each occupation was defined by the mean rating that employees have assigned to the scale required level of "thinking creatively". In studies of working conditions, the O*NET system represents an assessment principle labelled job exposure matrix. This particular normative "job exposure matrix" is based on a large sample of US employees provided by O*NET. Musicians ($n = 43$) were excluded.

Multivariate analyses (multiple linear regressions) were performed separately for men and women because we suspected that the associations might be partly gender-specific. First of all, alexithymia is more common in men than in women. Secondly, the female and male labour markets are different. The explanatory variables in our equations were the three aspects of musicality as well as age, education, general intelligence (WMT; see above) and alexithymia.

In the final step which included all the predictors, the results showed that among men ($n = 1327$) the variables that had independent statistically significant explanatory value were in falling order of power (the size of relative linear regression coefficients) education, cognitive ability, age and pitch discrimination ability. Alexithymia and rhythm and melody score did not have independent statistically significant predictive value.

For women ($n = 1908$), education, age, cognitive ability and (low) alexithymia score were the independent predictors of working in a creative occupation – in falling order of importance. In addition among women, a high rhythm score was a significant independent predictor before cognitive ability was included. But because of the strong correlation between WMT and rhythm score, the independent contribution of rhythm score became non-significant when WMT was included as a predictor, while the contribution of alexithymia remained significant.

From these results, we concluded that high-pitch discrimination ability is associated with the likelihood of having an occupation that requires creative thinking for men, regardless of education and general intelligence. The ability to handle emotions is associated with a creative occupation for women. As in the studies described above, the direction of causality cannot be established in this kind of study. Since subjects working professionally as musicians were excluded from the analyses, the associations are not confined to musical creativity but seem to have a more general nature. It is interesting that the patterns are different for men and women. We do not know whether this gender difference is due to the female/male difference in music activities, gender differences in levels of alexithymia or differences in the nature of the female and male labour markets, with more women working in health care and education and more men working in transportation and manufacturing, for instance.

It has to be pointed out that our study is cross-sectional. This means that we do not know whether the correlations we observe are due to differences in recruitment to or "survival" (staying) in creative occupations. In addition, both alexithymia and musicality could theoretically be influenced by exposure to work in creative occupations – which could give rise to reversed causation. The fact that musicality, alexithymia and general intelligence are interrelated further complicates the interpretation of our findings.

Conclusion

Our studies based upon a large cohort of twins from the Swedish Twin Registry have shown that musicality and ability to handle emotions are interrelated and that these variables are associated with likelihood of working in creative occupations. The patterns are different for men and women. While good ability to discriminate high from low notes (pitch) was independently predictive of work in creative occupations, a low alexithymia score was predictive of working in creative occupations in women. There was a pattern of gender differences. Boys started extra music lessons outside school less frequently than girls, but those boys who did start extra music were more likely to continue with music practice up to adult years. On the whole, the level of alexithymia assessed by means of the TAS 20 was higher in men than in women.

The relationships that we observe are of a small magnitude but may be important on a population basis. The extent to which ability to handle emotions can be affected by music interventions and education remains to be established. The relationship between extent of music practice and musical achievement on one hand and alexithymia on the other hand is determined to a substantial extent by genetic factors, and the importance of gene-environment interaction is probably great. A good society's obligation to provide artistic developmental opportunities for subjects with artistic drive in our view becomes obvious. There is an interplay between artistic activities and ability to handle emotions in relation to creativity in our society.

References

Bagby, R. M., Parker, J. D., & Taylor, G. J. (1994). The twenty-item Toronto alexithymia scale-I. Item selection and cross-validation of the factor structure. *Journal of Psychosomatic Research, 38*(1), 23–32.

Baumgartner, T., Lutz, K., Schmidt, C. F., & Jancke, L. (2006). The emotional power of music: How music enhances the feeling of affective pictures. *Brain Research, 1075*, 151–164. https://doi.org/10.1016/j.brainres.2005.12.065.

Bojner Horwitz, E., Lennartsson, A. K., Theorell, T. P., & Ullén, F. (2015). Engagement in dance is associated with emotional competence in interplay with others. *Frontiers of Psychology, 6*, 1096. https://doi.org/10.3389/fpsyg.2015.01096.

Borsci, G., Boccardi, M., Rossi, R., Rossi, G., Perez, J., Bonetti, M., et al. (2009). Alexithymia in healthy women: A brain morphology study. *Journal of Affective Disorders, 114*(1–3), 208–215. https://doi.org/10.1016/j.jad.2008.07.013.

Carson, S. H., Peterson, J. B., & Higgins, D. M. (2005). Reliability, validity, and factor structure of the creative achievement questionnaire. *Creativity Research Journal, 17, 37–50.* https://doi.org/10.1207/s15326934crj1701_4.

De Gucht, V., & Heiser, W. (2003). Alexithymia and somatisation: Quantitative review of the literature. *Journal of Psychosomatic Research, 54*(5), 425–434.

Deng, Y., Ma, X., & Tang, Q. (2013). Brain response during visual emotional processing: An fMRI study of alexithymia. *Psychiatry Research, 213*(3), 225–229. https://doi.org/10.1016/j.pscychresns.2013.03.007.

Dissanayake, E. (2000). Antecedents of the temporal arts in early mother-infant interaction. In N. L. Wallin, B. Merker, & S. Brown (Eds.), *The origins of music*. Boston: MIT Press.
Duan, X., Dai, Q., Gong, Q., & Chen, H. (2010). Neural mechanism of unconscious perception of surprised facial expression. *NeuroImage, 52*(1), 401–407. https://doi.org/10.1016/j.neuroimage.2010.04.021.
Eriksson, H., Harmat, L., Theorell, T., & Ullén, F. (2016). Similar but different: Interviewing monozygotic twins discordant for musical practice. *Musicae Scientiae*, 1–17. https://doi.org/10.1177/1029864916649791.
Fukunishi, I., Sei, H., Morita, Y., & Rahe, R. H. (1999). Sympathetic activity in alexithymics with mother's low care. *Journal of Psychosomatic Research, 46*(6), 579–589.
Gabrielsson, A. (2011). *Strong experiences with music – music is much more than just music*. Oxford: Oxford University Press.
Grabe, H. J., Schwahn, C., Barnow, S., Spitzer, C., John, U., Freyberger, H. J., et al. (2010). Alexithymia, hypertension, and subclinical atherosclerosis in the general population. *Journal of Psychosomatic Research, 68*, 139–147. https://doi.org/10.1016/j.jpsychores.2009.07.015.
Grape Viding, C., Osika, W., Theorell, T., Kowalski, J., Hallqvist, J., & Bojner Horwitz, E. (2015). The "culture palette"- a randomized intervention study for women with burnout symptoms in Sweden. *British Journal of Medical Practitioners, 8*(2), a813.
Heinzel, A., Schafer, R., Muller, H. W., Schieffer, A., Ingenhag, A., Eickhoff, S. B., et al. (2010). Increased activation of the supragenual anterior cingulate cortex during visual emotional processing in male subjects with high degrees of alexithymia: An event-related fMRI study. *Psychotherapy and Psychosomatics, 79*(6), 363–370. https://doi.org/10.1159/000320121.
Honkalampi, K., Hintikka, J., Saarinen, P., Lehtonen, J., & Viinamaki, H. (2000). Is alexithymia a permanent feature in depressed patients? Results from a 6-month follow-up study. *Psychotherapy and Psychosomatics, 69*(6), 303–308. https://doi.org/10.1159/000012412.
Jorgensen, R. S., & Houston, B. K. (1986). Family history of hypertension, personality patterns, and cardiovascular reactivity to stress. *Psychosomatic Medicine, 48*(1–2), 102–117.
Jorgensen, M. M., Zachariae, R., Skytthe, A., & Kyvik, K. (2007). Genetic and environmental factors in alexithymia: a population-based study of 8,785 Danish twin pairs. *Psychotherapy and psychosomatics, 76*, 369–375. https://doi.org/10.1159/000107565.
Juslin, P. N., & Sloboda, J. A. (Eds.). (2010). *Music and emotion: Theory and research*. Oxford: Oxford University Press.
Laiho, S. (2004). The psychological functions of music in adolescence. *Nordic Journal of Music Therapy, 13*(1), 47–63.
Laricchiuta, D., Petrosini, L., Picerni, E., Cutuli, D., Iorio, M., & Chiapponi, et al. (2014). The embodied emotion in cerebellum: A neuroimaging study of alexithymia. *Brain Structure and Function*. https://doi.org/10.1007/s00429-014-0790-0.
Lennartsson, A.-K., Bojner Horwitz, E., Theorell, T., & Ullén, F. (2017). Creative artistic achievement is related to lower levels of alexithymia. *Creativity Research Journal, 29*(1), 29–36.
Li, S., Zhang, B., Guo, Y., & Zhang, J. (2015). The association between alexithymia as assessed by the 20-item Toronto alexithymia scale and depression: A meta-analysis. *Psychiatry Research*. https://doi.org/10.1016/j.psychres.2015.02.006.
Luminet, O., Bagby, R. M., & Taylor, G. J. (2001). An evaluation of the absolute and relative stability of alexithymia in patients with major depression. *Psychotherapy and Psychosomatics, 70*(5), 254–260. https://doi.org/10.1159/000056263.
Madison, G. (2011). Cause and affect: A functional perspective on music and emotion. In F. Bacci & D. Melcher (Eds.), *Art and the senses* (pp. 329–350). Oxford: Oxford University Press.
Mantani, T., Okamoto, Y., Shirao, N., Okada, G., & Yamawaki, S. (2005). Reduced activation of posterior cingulate cortex during imagery in subjects with high degrees of alexithymia: A functional magnetic resonance imaging study. *Biological Psychiatry, 57*(9), 982–990. https://doi.org/10.1016/j.biopsych.2005.01.047.
Mattila, A. K., Saarni, S. I., Salminen, J. K., Huhtala, H., Sintonen, H., & Joukamaa, M. (2009). Alexithymia and health-related quality of life in a general population. *Psychosomatics, 50*(1), 59–68. https://doi.org/10.1176/appi.psy.50.1.59.

Mattila, A. K., Saarni, S. I., Alanen, E., Salminen, J. K., Kronholm, E., Jula, A., et al. (2010). Health-related quality-of-life profiles in nonalexithymic and alexithymic subjects from general population. *Journal of Psychosomatic Research, 68*(3), 279–283. https://doi.org/10.1016/j.jpsychores.2009.09.010.

Moriguchi, Y., Ohnishi, T., Lane, R. D., Maeda, M., Mori, T., Nemoto, K., et al. (2006). Impaired self-awareness and theory of mind: An fMRI study of mentalizing in alexithymia. *NeuroImage, 32*(3), 1472–1482. https://doi.org/10.1016/j.neuroimage.2006.04.186.

Nemiah, J. C. (1996). Alexithymia: Present, past–and future? *Psychosomatic Medicine, 58*, 217–218. https://doi.org/10.1097/00006842-199605000-00004.

Porges, S., & Byrne, E. (2012). Research methods for measurement of heart rate and respiration. *Biological Psychology, 34*(2–3), 93–130. http://www.sciencedirect.com/science/article/pii/030105119290012J. Accessed June 25, 2012.

Ruud, E. (2010). *Music therapy: A perspective from the humanities*. Gilsum: Barcelona Publishers.

Salminen, J. K., Saarijarvi, S., Aairela, E., & Tamminen, T. (1994). Alexithymia–state or trait? One-year follow-up study of general hospital psychiatric consultation out-patients. *Journal of Psychosomatic Research, 38*(7), 681–685.

Salminen, J. K., Saarijarvi, S., Aarela, E., Toikka, T., & Kauhanen, J. (1999). Prevalence of alexithymia and its association with sociodemographic variables in the general population of Finland. *Journal of Psychosomatic Research, 46*(1), 75–82.

Sifneos, P. E. (1973). The prevalence of 'alexithymic' characteristics in psychosomatic patients. *Psychotherapy and Psychosomatics, 22*(2), 255–262.

Sifneos, P. E. (1996). Alexithymia: Past and present. *American Journal of Psychiatry, 153*(Suppl. 7), 137–114.

Theorell, T. (2014). Psychological health effects of musical experiences. In *Theories, studies and reflections in music health science*. Dordrecht: Springer.

Theorell, T. P., Lennartsson, A. K., Mosing, M. A., & Ullén, F. (2014). Musical activity and emotional competence – a twin study. *Frontiers in Psychology, 5*, 774. https://doi.org/10.3389/fpsyg.2014.00774.

Theorell, T., Lennartsson, A. K., Madison, G., Mosing, M. A., & Ullén, F. (2015). Predictors of continued playing or singing – from childhood and adolescence to adult years. *Acta Paediatrica, 104*(3), 274–284. https://doi.org/10.1111/apa.12870.

Theorell, T., Madison, G., & Ullén, F. (2017). Associations between alexithymia, musical ability, and working in a creative occupation: Different patterns for men and women. Submitted 2017.

Tolmunen, T., Heliste, M., Lehto, S. M., Hintikka, J., Honkalampi, K., & Kauhanen, J. (2011). Stability of alexithymia in the general population: An 11-year follow-up. *Comprehensive Psychiatry, 52*(5), 536–541. https://doi.org/10.1016/j.comppsych.2010.09.007.

Ullén, F., Mosing, M. A., Holm, L., Eriksson, H., & Madison, G. (2014). Psychometric properties and heritability of a new online test for musicality, the Swedish musical discrimination test. *Personality and Individual Differences, 63*, 87–93.

Ullén, F., Hambrick, D. Z., & Mosing, M. A. (2016). Rethinking expertise: A multifactorial gene-environment interaction model of expert performance. *Psychological Bulletin, 142*(4), 427–446. https://doi.org/10.1037/bul0000033.

Part II
Theoretical Perspectives on Music and Public Health

Chapter 5
How Does Music Translate Itself Biologically in a Public Health Context?

Töres Theorell

Introduction

This chapter is devoted mainly to bodily concomitants of musical experiences with focus below the brain. Several reviews have been published in this field. A more detailed analysis can be found in Theorell (2014). In the present chapter, I have no intention to give a full account. I have selected a few examples, which intend to show how researchers are thinking about 'translation' of neurobiological brain processes to bodily reactions. Most of these examples deal with emotionally and socially charged stimuli.

Of course all psychological processing starts in the brain, but the body is an important part in the total experience. Le Doux (1998) and co-workers described emotionally charged stimuli to be transmitted to the brain via the two contrasting 'upper' and 'lower' routes. These routes could also be labelled 'slower' and 'faster'. When we hear a piece of music which has a strong emotional meaning for us, the impulse reaches the thalamus located in the midbrain – a relay station that directs stimuli to other parts of the brain. The faster route transmits the emotional information to the emotional brain. This part of the brain has an important role in stress and anxiety reactions but also in the regulation of other emotions such as sadness, anger and joy. The emotional brain rapidly triggers a response corresponding to the emotion in the brain and the rest of the body. The impulse is also carried via the slower route up to the brain cortex which activates cognitive interpretation. Since this route is much slower, the cognitive brain may never become aware of a rapidly disappearing stimulus. In such a case, it is only the lower faster route that has been activated, and the brain may never become conscious of the source of the emotional reaction. Despite this, the cognitive processes may be influenced by these emotional impulses.

T. Theorell (✉)
Department of Neuroscience, Karolinska Institute, Stress Research Institute, Stockholm University, Stockholm, Sweden
e-mail: tores.theorell@su.se

L. O. Bonde, T. Theorell (eds.), *Music and Public Health*,
https://doi.org/10.1007/978-3-319-76240-1_5

The cognitive brain is 'surprised' and may start analysing life problems in new unexpected ways. This kind of brain surprise may be an important reason why music therapy and social uses of music could add something to logical conversations in rehabilitation or in treatments of depression (for a discussion, see Theorell 2014).

Neurobiological Background: Emotional and Social Competence

It has been known for a long time that different emotions correspond to different specific combinations of psychophysiological reactions in the body. Already in 1997 a team of researchers (Krumhansl 1997) described how a group of individuals reacted physiologically to three different kinds of musical pieces that were assumed to induce *sadness* (Albinoni, *Adagio*; Barber, *Adagio*), *fear* (Mussorgsky, *Witch Hill*; Holst, 'Mars', *The Planets*) and *joy* (Vivaldi, *Spring* from *The Four Seasons*; Alfvén, *Midsummer Wake*), respectively. It was shown that the music really induced the expected feelings in most experimental subjects. Secondly, it turned out that the physiological reactions were indeed different in these induced emotional states. During the sad music, the persons had lower heart rate, higher blood pressure and more sweating (galvanic skin response). With fearful music there was increased respiratory rate (number of breaths per minute) and decreased blood flow in the periphery (finger tips). The joyful music was associated mainly with an increased respiratory rate. How specific music can amplify group feelings, especially depressed ones, has been examined by Garrido et al. (2017).

An important concept in the psychophysiological understanding of possible beneficial health effects of cultural experiences is 'emotional competence' (see, for instance, Morris et al. 1998; Juslin and Timmers 2010; Theorell 2014). This refers to the fact that emotional impulses and rational understanding are handled differently by different parts of the brain and that the cooperation of those is vital to health. In psychosomatic medicine the concept *alexithymia* is important (see Chap. 4 by Theorell and Ullén). Alexithymia means lack of ability to differentiate feelings (Sifneos 1996). A person who has a good ability to identify, describe, differentiate and deal with different emotions has a great advantage. By contrast those who suffer from alexithymia have poor such ability. Our emotions serve as a driving force and a compass in many situations. Biologically speaking we show *anger* because we want others to be afraid of us. *Sadness*, particularly when accompanied by tears, shows that we are in need of help (Hendriks et al. 2008). We show *pride* because we expect the environment to praise us. This feeling can also amplify a positive collective feeling. All our emotions serve as signals to the environment, and at the same time, they are driving forces in our own acts. Every emotion has both bodily and psychological expressions. A person who has good skills in differentiating feelings is also likely to be able to differentiate feelings in relatives, friends and family members (see Baughman et al. 2013). Accordingly emotional competence relates to social competence. There is a separate chapter on emotion handling in this book (see Chap. 4 by Theorell and Ullén; see also Chap. 7 by Saarikallio and Baltazar).

Alexithymia is associated with increased illness risk; the most well-documented example is hypertension (high blood pressure), which arises more often in alexithymic persons than in others (Grabe et al. 2010). As pointed out in the chapter on alexithymia, about 30% of this is determined by genetic factors. In addition the *association* that we have shown between extensive music practicing and improved emotional skill is mainly genetically determined also. Therefore we do not know to what extent early music training can improve emotional skills. However, a controlled experiment performed by our group on fifth and sixth graders in a public school in Stockholm (Lindblad et al. 2007) showed that saliva cortisol levels during school hours were lowered (indicating lowered stress levels in the class room) when a special extra music hour was introduced every week aiming at stimulated togetherness in the class room. No such improvements were observed in two comparison groups. The difference between the groups was not observable until the end of the school year indicating that these favourable changes in physiological stress patterns evolved slowly during the experimental year. The tentative interpretation of this is that the children in the music group developed better social skills and that they were able to feel with the other children. Obviously this could lead to an improved environment for learning. A much larger experiment was performed in Switzerland some years ago (see Spychiger 2002). This comprised 52 classes distributed across the whole country. Half of these classes were randomly assigned to the music group. One hour every week of 'social' (ensemble) music training was introduced for 3 years. Standardized measurements showed that the music classes developed their social skills better than did the children in the comparison group. In addition, language and mathematics skills developed as well or even better in the music group despite the fact that the introduction of the extra music lessons reduced the number of language and mathematics lessons slightly. Parent singing with newborn children plays an important role during the first weeks of life (Dissanayake 2000). More intervention studies are needed. In addition we do not know how early we should start. A qualified guess, however, is that we should start already in the foetal stage! When the foetus is in its third trimester of the pregnancy, it has fairly good hearing already, and it is able to differentiate sounds (Granier-Deferre et al. 2011). It has also been shown that the foetus can learn to differentiate vowels that are audible for it during this part of the pregnancy – the child shows physiological signs (sucking behaviour) indicating that it recognizes the mother's speech sounds from the foetus period when it is newborn (Moon et al. 2013). An American newborn child recognizes typical American speech sounds, while a Swedish child recognizes typical Swedish speech sounds. Speech melody – so-called prosody – is close to music for the child!

Music Psychophysiology

Gabrielsson (2011) recruited more than 900 participants for an interview study. He asked them to describe in their own words the most profound musical experience that they had had in their lives and then went on to categorize these experiences with regard to contents, context and consequences. Gabrielsson stated that it was very difficult to do such a categorisation since the experiences cover a very large area, asserting that music seems to comprise the 'whole psychological reality'. Many of these situations were seen as being turning points in people's lives. He provided examples of people who reported that they had been deeply depressed when they discovered a kind of music that they had never been interested in before. In their depressed mood, they became passionately engaged in listening to this particular kind of music. In retrospect they believed that this experience helped them out of the depression. These descriptions may illustrate the point above regarding the 'surprised brain'. A strong acute reaction may have profound long-term effects on health since a reorientation in life may take place as a result.

That music may have immediate effects on several organ systems in the body is becoming increasingly clear with a rising number of recording equipment for different organ systems. For instance, recordings of peristaltic movements in the stomach (Chen et al. 2005) performed by means of a pressure recording device swallowed by the experiment person have shown that harmonic slow classical music stimulated well-coordinated slow peristaltic movements (supposedly good for alimentation!), whereas noise stimulated fibrillating uncoordinated peristaltics. More indirectly, assessments of the ability of the arterial wall to recuperate after compression have been performed (Miller et al. 2010). The experiment showed that subjects who listen to happy music while an artery in the forearm is compressed (as in blood pressure assessment) have a better recuperation afterwards (when the compression was released) compared to when they had listen to anxiety-provoking music. This relates to the function of the endothelial cells on the inside of the artery and could theoretically be of significance in atherosclerosis.

Music is being used increasingly in medical settings, for instance, for pain management (for a review, see Bernatzky et al. 2011) but also in treatment of depression, after surgery and other painful medical procedures, dementia care and Parkinson's disease, just to mention a few examples (see Theorell 2014; Jensen and Bonde 2018). Piloerection is an interesting phenomenon, which arises unexpectedly when subjects listen to music and become surprised by the music's strong effect. Piloerection is literally erection of hair. All fur animals have the ability to erect the straws in their fur, either because they need to increase their size or in order to increase isolation against cold temperature. In an experiment (Vickhoff et al. 2012) that we performed, the listening subject had an unexpected piloerection when piano music was improvised during a change from calm, slow, rhythmic and harmonic music to disharmonic rhythm with irregular beat. Physiological recordings showed at first increasing heart rate and at the same time diminished heart rate variability, thereafter increased sweating (skin conductance) and finally decreased finger temperature (decrease of 0.7 °C during 40 s). The cold skin temperature coincided with piloerection. Again this illustrates how strong our physiological reaction can be to musical stimuli, a well-known observation among music therapists!

In an experiment performed on a group of 37 young adults, the participants were asked to select two of their own favourite music pieces (Lingham and Theorell 2009), one piece that according to their own assessment was 'stimulating' and one that they considered 'relaxing'. While sitting quietly they listened in random order to these pieces. Their heart rate, breathing frequency and emotional state were recorded, and recordings during listening to the favourite pieces were compared to preceding restful silence. During the stimulating piece, the average increase in heart rate was seven beats per minute compared to the quiet condition. The relaxing music on the other hand in some individuals did not produce the expected deceleration of heart rate. Quite to the contrary, in 11 cases, it was even associated with a small increase in heart rate. There were similar observations with regard to breathing. During the stimulating music, the breathing frequency increased significantly, and the average increase was four breathing cycles per minute. During relaxing music on the other hand, there was no significant change. Emotional self-recordings verified this in the psychological domain; whereas the stimulating music quite clearly induced arousal feelings, the relaxing piece triggered both relaxed and aroused feelings. The first conclusion to be drawn from this seems to be that these highly educated people were skilful in selecting their own stimulating music, but they were not as successful with regard to relaxing music – it frequently did not have the expected calming effect on heart rate and breathing. The second conclusion is that responses (even) to self-selected music showed enormous interindividual variation. Despite the fact that the subjects did not move some of them had strong heart rate reactions, one subject's heart rate increased by 20 beats per minute while she was listening to the stimulating music and decreased by 20 beats when she listened to the relaxing music. Other participants' heart rates did not change at all during music listening.

Research has shown that specially adapted music may decrease both the subjective experience of physical effort and the physiological reactions to it. Szmedra and Bacharach (1998), for instance, did such an experiment several years ago. Young men were asked to do exactly the same physical work on the treadmill during two different conditions. One condition was in silence and the other was while listening to special gym music. The two conditions were randomly ordered. The blood concentration of lactic acid was assessed. Heart rate and blood pressure were recorded, and the subjects were asked to make a self-rating of effort. Despite the fact that the physical work was fixed and exactly the same in the two conditions, blood pressure and heart rate elevation was smaller during the music condition, and the same observation was made for subjective effort and lactic acid concentration.

Other research in this field has shown that in long-distance cycling (10 km), when the cyclists choose their own tempo, listening to fast dance music ('trance' with tempo 142 beats per minute) is likely to induce an elevated cycling tempo and the work being perceived as harder without music (Atkinson et al. 2004). In another study the effects of different types of music on heart rate, rating of perceived exertion and time to exhaustion were studied. The subjects performed their physical work on a treadmill, and the conditions were randomly allocated to 'soft music'; 'loud, fast and exciting music'; and silence. The results showed that the soft music reduced physiological and psychological arousal during submaximal exercise and also increased endurance of performance (Copeland and Franks 1991).

In many activities typical to the modern world, music is used in ways that may seem new but are in fact very old. A striking example is the music played in the gym. Special kinds of music have been developed for this (DeNora 2000). One is reminded of all music that has been used in the history of man in order to facilitate physical work, for instance, 'pulling boat' songs and sailor songs for rowing or managing big sails and march music for facilitating long troop walks (see also Merriam (1964)).

Psychophysiology of Music Performance

Human music performance requires physical and mental effort. Hundreds of muscles are engaged during singing and also during playing instruments. Singing in groups frequently entails bodily movement. Even listening can induce strong physical activations, for instance, when we are moving to dance music or attending a rock concert. Therefore, enhancement of aerobic skill and fitness may be inherent to large parts of musical experience. Vice versa, while performing aerobic exercises, music can act as a motivating stimulus. Some of these effects appear to be exploited in commercial dance and gymnastic studios. In another vein, empirical studies of professional and amateur singers suggest that learning to sing in a controlled way can be associated with a systematic training of breathing technique. Professional singers differ from amateurs with regard to heart rate variability (HRV) during singing (Grape et al. 2003) – the professionals show much more heart rate variability than amateurs. Most of the heart rate variability is determined by activity in the parasympathetic nervous system that in many ways counterbalances the sympathetic nervous 'stress' system. During singing the professionals manage to activate a high degree of heart rate variability. This may be due to a trained breathing technique with emphasis on diaphragmal breathing. Diaphragm breathing stimulates the vagus nerve – the most important component in the parasympathetic nervous system. Breathing techniques acquired through musical practice could be used in situations outside the music room as well, producing a 'transfer' effect – from music performance to an advantage in everyday non-singing life. Clift and co-workers have studied the benefits of choir singing in large clinically and nonclinically defined groups of choir singers (Clift et al. 2010, Clift 2012). Lewis et al. (2016) have reviewed the literature on the potential benefits of choir singing for maintaining lung health among subjects with chronic respiratory disorders. The literature points out, for instance, that choir singing improves the ability of the singer to maximize the emptying of the lungs.

Sometimes stage performers can experience 'flow' (often referred to as 'effortless attention'; see Czikszentmihalyi 1990; Ullén et al. 2010). The state of flow typically arises when a person is performing a difficult task but has a sufficient skill level to match the task challenge. Characteristic components of the flow experience are high but subjectively effortless attention, a sense of control, low self-awareness, a distorted sense of time and positive affect (Csikszentmihalyi and Nakamura 2010). This state may arise in many activities and is in no way confined to stage performances. Flow experiences can elevate quality of life. An online study (De Manzano et al. 2010) of physiological states in professional pianists when they played a difficult piece (which they liked and had selected) on five different occasions with varying degree of flow (according to a standardized questionnaire-based assessment) showed that the flow state was characterized by a high state of arousal (with high heart rate), increased activity in the 'smile muscle' and increased breathing depth. The increased breathing depth is likely to have increased the parasympathetic activity mediated by the vagus nerve. One possibility is therefore that flow, in physiological terms, is a state of high arousal, which is counterbalanced by high activation of the vagus nerve. According to the 'polyvagal theory' introduced by Porges (2007), this balance between sympathetic and parasympathetic activity is influenced by previous experiences (for instance, traumatic events), and it can be trained and can be favourably affected by musical experiences. Clearly more research is needed on the physiological underpinnings of flow in various activities, however.

Immediate effects of singing have been shown in several studies. For instance, choir singers have been reported to have an increased saliva content of immunoglobulins after a rehearsal compared to before the rehearsal (Kreutz et al. 2004). A study of the immediate effects of the individual singing lesson (Grape et al. 2003) showed that singing – both in amateurs and in professionals – was associated with an increased plasma concentration of oxytocin. This may be a non-specific reaction to a situation with a strong social interaction component. A more recent experiment (Kreutz 2014) showed that the saliva concentration of oxytocin increased in choir singers after 30 min of rehearsing, but it did not increase in the same persons after 30 min of group conversation. Oxytocin seems to play a role in social bonding. Therefore emotions related to the coherence in the group were also measured on the same occasions. These were standardized pre-post assessments of 'positive' and 'negative' emotions. These showed that the emotions became significantly more positive and less negative after choir singing but that there were no changes after the conversation. It could be speculated that the frequent repetition of active singing with all its hormonal effects could give rise to an accumulated wellbeing effect over time. However, the significance in the long run of this kind of effect on health is unknown.

Studies of Long-Term Bodily Effects of Repeated Music Making or Dancing

What are the possible mechanisms mediating health-promoting effects of music in adult participants? Most of the research of biological correlates of music experiences has been based upon experiments of a short duration. There are few convincing published studies on long-term health effects of repeated music experiences. Such studies should ideally be performed on groups that are genuinely comparable from start and who are followed over time. A study of elderly institutionalized subjects in Washington DC (Cohen 2009) filled these criteria. Participants were recruited among those who wanted to start singing in a choir, but after randomization only half were allowed to do so, whereas the others constituted the comparison group and had to wait for organized singing during 2 years. According to standardized questionnaire-based assessment, the singing participants had a significantly better health development during the 2 years than those in the comparison group. New large-scale studies of this nature are under way in the world.

Our own research group performed a randomized small-scale study of patients with irritable bowel syndrome (IBS), a psychosomatic bowel disease with symptoms that are aggravated by worsened life conditions. It is common in the population (approximately 10%). We assessed the saliva excretion of testosterone (Grape et al. 2010) and the blood concentration of a few biochemical indicators (Grape et al. 2009). None of the participants had been singing in a choir but wanted to do so. A randomly selected half of the patients were now offered the possibility to sing in a choir once a week during a year. Those randomly allocated to the control group had IBS-related group activities (lectures and group discussions) without singing once a week during a year. Assessments were made on all participants before start and then 6, 9 and 12 months later. Saliva testosterone was assessed on six occasions during the wake hours of the four measurement days. Testosterone is important both in men and women for the regeneration of cells and for the body's protection against stress-related disorders. Variations in the concentration of saliva reflect variations in blood concentration and hence in regenerative activity.

The findings in the choir study showed that after 6 months, there was a pronounced increase in the excretion of testosterone in the choir group but not in the comparison group. After 1 year the differences were not statistically significant any more. Blood samples were collected before start and 1 year later. Plasma fibrinogen concentration, an indicator of long-lasting arousal, was assessed in these samples – the findings indicating a favourable 1-year effect on the choir group relative to the other group (Grape et al. 2009). Our findings also showed that to the advantage of the choir group during the study year, there was an almost significant ($p < 0.10$) improvement of gut-related pain and plasma concentration of *motilin*, a hormone that regulates bowel movements and is important for this group of patients. Compared to the control group, there was also a significantly better development of fibrinogen concentration in the patient group during the study year. The results indicate that group singing may stimulate regeneration and reduce long-term stress. The results show, however, that the effects of choir singing may depend on context – the choir group may not have functioned as well during the end of the year as it did during the first half-year. These findings illustrate physiological mechanisms that may explain perceived health-promoting effects of group singing. Much larger studies of representative samples are required for sound conclusions.

Researchers have also examined to what extent biological functions may be synchronized between singers in a choir (Vickhoff et al. 2013; Olsson et al. 2013). There is evidence that not only breathing patterns but also heart variability may become synchronized between singers in a choir. However, this is when the singers are doing the music without words slowly without complicated rhythm and without complicated differences in vocal parts (separately for sopranos, altos, tenors and basses). During humming and simple singing of hymns, there is synchronization of respiration and heart rate variability.

Studies of the collective beneficial effects of choir singing are rare, but such studies have been published. Coulton et al. (2015) performed a large randomized trial on elderly in a community singing study. After 6 months there were significant beneficial effects on mental health according to standardized measures based upon SF 12, a standardized questionnaire widely used for the assessment of mental health.

Dance and Music

In their research Krantz et al. (2006) have shown that lay people in Sweden react in a predictable way when diatonic ascending chords (defined as two notes at the same time) are being played. Minor second (two notes as close as possible to one another on a 12-tone scale, for instance, c and c sharp) and major seventh (two notes that are as almost and as close as possible to an octave, for instance, A and a flat) are the most disharmonic diatonic chords. When these chords are being played, the experimental persons tend to report disharmonic emotions, such as worry, irritation and sadness. Chords in the middle range tend to trigger soft or agreeable feelings. The major *sixth* was associated with the most pronounced *joy*, whereas the *fifth* was associated with *wholeness*, the major *third* with *embracing* and the major *seventh* with *uncoordinated jittery* movements. The fact that specific movements are coupled with specific diatonic chords indicates that dance and music may be programmed together in our brains. This may have had phylogenetic meaning. In addition it was shown (Krantz et al. 2010) that 'naïve' listeners (i.e. not trained in music or dancing) who were exposed randomly either to the major third or to the major seventh had strikingly different cardiovascular reactions to these two different diatonic chords. The difference was particularly evident when the subject heard the chord for the first time during the experiment, and it disappeared when the participant was allowed to move (in comparison to no movement). The disharmonic chord induced a short period of irregular heartbeat, whereas no such reaction was observed in the other situation.

In a randomized trial study of 36 patients, Bojner Horwitz and co-workers (2003) have evaluated the effects of dance therapy for patients with fibromyalgia. A novel procedure in that research was film recording of movement patterns in standardized situations. Such recordings were performed before the dance therapy started as well as 14 months later. After 14 months standardized ratings of the patients' movements showed that the movement patterns had improved in the dance group compared to the control group. Self-ratings and assessments of hormones did not show significant improvements. Patients in the dance group, however, tended to have higher concentration of cortisol in plasma and saliva after therapy compared to before – as a possible indicator of activation of the hypothalamo-pituitary-adrenocortical (HPA) axis. The HPA axis is instrumental in energy mobilization in stressful situations so increased HPA activity in healthy subjects is an indicator of increased energy mobilization – stress. However, inability to activate the HPA axis is frequently a component in long-lasting physiological exhaustion. An increased cortisol level in patients who have shown clinical symptoms of exhaustion (as patients with fibromyalgia do) could be interpreted positively because of this known association between chronic stress conditions and passiveness in the HPA axis. This study illustrates some of the methodological difficulties in quantitative scientific studies of the effects of dance. The patients are often unaware of their own improvement in movement patterns, and several effects may be associated with activation of hormonal systems rather than soothing of hyperactive patterns.

Dancing may also activate anabolic/regenerative hormones. A study by Quiroga et al. (2009) of the hormonal effects of dancing Tango Argentino showed that during tango dancing, when there is a partner and when music is being played, testosterone excretion is stimulated both in women and men. A randomized study has been published by Kattenstroth et al. (2013) on the effects of dance class once a week during 6 months on elderly women and men (age 60–94) with 25 subjects in the intervention and 10 in the control group. Significant improvements in lifestyle index and tests of cognition, reaction time, posture, hand/motor function and tactile ability were found after the intervention period in the experimental group but not in the control group. These promising results show that dance class may be a very beneficial activity for elderly, especially demented subjects (see, for instance, Jensen and Bonde 2018, Bonde and Ochsner Ridder 2017). A crucial factor is that dancing may facilitate activation of memories, and this could facilitate daily care of these demented patients (see Hammar et al. 2010). Bittman et al. (2003) have also shown that recreational music making with demented patients in institutions can improve the work environment for staff and reduce turnover. Free dance has also been used successfully for teenage girls with depressive and psychosomatic symptoms. A randomized control study (Duberg 2016) of the effects of free dance instruction once a week for 8 months showed beneficial effects on symptom development in the intervention group. After 20 months the differences between the groups were not significant – perhaps an indication that the dancing sessions have to continue! This is of course an observation that could be applied to many cultural activities.

These findings are consistent with functional magnetic resonance imaging studies (Pallesen et al. 2005), which show that simple diatonic chords (major, minor and dissonant) correspond to specific differential activation patterns in the brain.

It has been known for a long time in physiotherapy that a thorough analysis of patients' bodily language reveals a lot about their emotional state. This has been used extensively in dance therapy.

Art Psychotherapy in Rehabilitation

Sometimes several forms of cultural experiences may interact in generating forceful effects (e.g. dance and music, visual arts and music, music and words). For instance, there is controversy over the existence of music modules in the brain (i.e. brain regions that are specific to processing music). Very briefly, many neural structures and functions that are activated during music experiences are shared or overlap with other domains, such as emotion, movement (e.g. corresponding to correlates of finger or limb movements during listening to music) and speech and language. There is initial evidence showing that music can enhance activations of brain regions that are relevant in one type of activity and vice versa. For instance, Baumgartner et al. (2006) investigated brain responses to unimodal (picture of art or music) and bimodal (picture of art and music at the same time) presentations of images and music to healthy participants. Brain areas known to be involved in auditory as well as in neutral and emotional visual-auditory integration processes were activated in the respective picture and music conditions. In the combined condition, there was increased activation of many emotion-processing structures (e.g. amygdala and hippocampus). In contrast, the picture condition only showed an activation increase in the cognitive part of the cortex. These findings indicate that emotional pictures evoke a more cognitive mode of emotion perception, whereas congruent presentations of emotional visual and musical stimuli rather automatically evoke strong emotional feelings and experiences. It is evident that contextual factors can significantly modulate responses to musical activities in the brain. Another conclusion is that music experiences may amplify concomitant experiences in other modalities, which can be significant in a health context.

Theorell et al. (1998) performed a study on 24 patients who suffered from chronic pain and other psychosomatic conditions. They were all on sick leave and had been so for at least a year. These patients were followed for 2 years with blood samples and standardized questionnaires illuminating mental state every fourth to sixth month. Each patient was informed that she/he would be treated during a 2-year period. In the art psychotherapy programme, the patient was treated once a week. The patients were assigned according to anticipated needs and preferences to one of the programmes: visual art, music, dance and psychodrama. However, the group of therapists met once a month, and the patients' therapy courses were reviewed by the therapists together. Sometimes painting or drawing was used for the patients who did not participate in the visual art group. Similarly movements (dance), theatre experiences and music were used in the groups not specializing on these activities. Because of the longitudinal nature of the data collection, it was possible to follow typical time courses in these treatments. When the patients started, they were mostly in a low energy state (mirrored in a low concentration of uric acid in serum). When the therapist and patient started knowing one another, crucial art experiences during the therapy sessions could evoke memories of traumatic experiences. During this phase – mostly after approximately 6 months – the energy levels (serum uric acid as well as self-reported) were rising. At the end of the first year, most patients had calmed down and the energy level was intermediate. During this phase the levels of anxiety and depression had decreased. It lasted until 2 years after start before somatic symptoms tended to decrease however.

Conclusions

This chapter is devoted mainly to bodily concomitants of musical experiences with focus below the brain. We have seen that science has made efforts for a long time to understand the body's immediate translation of music experiences into physiological reactions. Immediate effects of music experiences are important to public health. Long-term effects, however, have been less extensively studied. The long-term effects are perhaps the most important ones from a public health point of view. That such effects have been less studied is understandable because it takes much more time to study long-term effects and it is difficult to construct reliable research designs. There are indications from available research that cultural activities and engagement have to continue for a long time in order to give long-term measurable bodily effects. On the other hand, it is also true that strong cultural experiences with such a peak experience during music listening could change the direction of a whole life. I have not made any effort to cover the rich literature. Instead I have given examples, which illustrate how researchers are thinking about music translation processes in the body.

References

Atkinson, G., Wilson, D., & Eubank, M. (2004). Effects of music on work-rate distribution during a cycling time trial. *International Journal of Sports Medicine, 25*(8), 611–615.

Baughman, H. M., Schermer, J. A., Veselka, L., Harris, J., & Vernon, P. A. (2013). A behavior genetic analysis of trait emotional intelligence and alexithymia: A replication. *Twin Research and Human Genetics, 16*(2), 554–559. https://doi.org/10.1017/thg.2012.151.

Baumgartner, T., Lutz, K., Schmidt, C. F., & Jäncke, L. (2006). The emotional power of music: How music enhances the feeling of affective pictures. *Brain Research, 1075*, 151–164.

Bernatzky, G., Presch, M., Anderson, M., Panksepp, J. (2011). Emotional foundations of music as a non-pharmacological pain management tool in modern medicine. *Neuroscience & Biobehavioral Reviews, 35*, (9), 1989–1999.

Bittman, B., Bruhn, K. T., Stevens, C., Westengard, J., & Umbach, P. O. (2003). Recreational music-making: A cost-effective group interdisciplinary strategy for reducing burnout and improving mood states in long-term care workers. *Advances in Mind-Body Medicine, 19*(3–4), 4–15.

Bojner Horwitz, E., Theorell, T., & Anderberg, U. (2003). Dance/movement therapy and changes in stress-related hormones: A study of fibromyalgia patients with video-interpretation. *The Arts in Psychotherapy, 30*, 255–264.

Bonde, L. O., & Ochsner Ridder, H. M. (2017). Erindringsdans og livets sange. Musik, sang og dans som oplevelse og oplivelse i demensomsorgen i Danmark. In A. Jensen (Ed.), *Kultur og sundhed*. København: Turbine.

Chen, D. D., Xu, X., Zhao, Q., Yin, J., Sallam, H., & Chen, J. D. (2005). Alteration of gastric myo-electrical and autonomic activities with audio stimulation in healthy humans. *Scandinavian Journal of Gastroenterology, 40*, 814–821.

Clift, S. M. (2012). Singing, wellbeing, and health. In R. Macdonald, G. Kreutz, & L. Mitchell (Eds.), *Music, health & wellbeing* (pp. 113–124). Oxford: Oxford University Press.

Clift, S., Hancox, G., Morrison, I., Hess, B., Kreutz, G., & Stewart, D. (2010). Choral singing and psychological wellbeing: Quantitative and qualitative findings from English choirs in a cross-national survey. *Journal of Applied Arts and Health, 1*(1), 19–34. https://doi.org/10.1386/jaah.1.1.19/1.

Cohen, G. (2009). New theories and research findings on the positive influence of music and art on health with ageing. *Arts & Health, 1*, 48–63.

Copeland, B. L., & Franks, B. D. (1991). Effects of types and intensities of background music on treadmill endurance. *The Journal of Sports Medicine and Physical Fitness, 31*(1), 100–103.

Coulton, S., Clift, S., Skingley, A., & Rodriguez, J. (2015). Effectiveness and cost-effectiveness of community singing on mental health-related quality of life of older people: Randomised controlled trial. *British Journal of Psychiatry, 207*(3), 250–255. https://doi.org/10.1192/bjp.bp.113.129908.

Csikszentmihalyi, M., & Nakamura, J. (2010). Effortless attention in everyday life: A systematic phenomenology. In B. Bruya (Ed.), *Effortless attention: A new perspective in the cognitive science of attention and action* (pp. 179–190). Cambridge, MA: The MIT Press.

Czikszentmihalyi, M. (1990). *Flow: The psychology of optimal experience.* New York: Harper and Row.

De Manzano, Ö., Harmat, L., Theorell, T., & Ullén, F. (2010). The psychophysiology of flow during piano playing. *Emotion, 10*(3), 301–311.

DeNora, T. (2000). *Music in everyday life.* Cambridge: Cambridge University Press.

Dissanayake, E. (2000). Antecedents of the temporal arts in early mother-infant interaction. In N. L. Wallin, B. Merker, & S. Brown (Eds.), *The origins of music.* Boston: MIT Press.

Duberg, A. (2016). *Dance intervention for adolescent girls with internalizing problems. Effects and experiences.* Doctoral thesis. Örebro Studies in Medicine nr 144, 2016.

Gabrielsson, A. (2011). *Strong experiences with music – Music is much more than just music.* Oxford: Oxford University Press.

Garrido, S., Eerola, T., & McFerran, K. (2017). Group rumination: Social interactions around music in people with depression. *Frontiers in Psychology, 8*, 490. https://doi.org/10.3389/fpsyg.2017.00490. eCollection 2017.

Grabe, H. J., Schwahn, C., Barnow, S., Spitzer, C., John, U., Freyberger, H. J., et al. (2010). Alexithymia, hypertension, and subclinical atherosclerosis in the general population. *Journal of Psychosomatic Research, 68*(2), 139–147.

Granier-Deferre, C., Ribeiro, A., Jacquet, A. Y., & Bassereau, S. (2011). Near-term fetuses process temporal features of speech. *Developmental Science, 14*(2), 336–352.

Grape, C., Sandgren, M., Hansson, L.-O., Ericson, M., & Theorell, T. (2003). Does singing promote well-being? An empirical study of professional and amateur singers during a singing lesson. *Integrative Physiological and Behavioral Science, 38*, 65–74.

Grape, C., Theorell, T., Wikström, B. M., & Ekman, R. (2009). Choir singing and fibrinogen, VEGF, cholecystokinin and motilin in IBS patients. *Medical Hypotheses, 72*, 223–225.

Grape, C., Wikström, B. M., Hasson, D., Ekman, R., & Theorell, T. (2010). Saliva testosterone increases in choir singer beginners. *Psychotherapy and Psychosomatics, 79*, 196–198.

Hammar, L. M., Emami, A., Engström, G., & Götell, E. (2010). Reactions of persons with dementia to caregivers singing in morning care situations. *The Open Nursing Journal, 4*, 35–41. https://doi.org/10.2174/1874434601004010035.

Hendriks, M. C., Croon, M. A., & Vingerhoets, A. J. (2008). Social reactions to adult crying: The help-soliciting function of tears. *The Journal of Social Psychology, 148*(1), 22–41. https://doi.org/10.3200/SOCP.148.1.22-42.

Jensen, A., & Bonde, L.O. (2018). Deltagelse i kunst- og kulturaktiviteter og kunstbaserede interventioner have positiv effekt på somatiske sygdomme? *Ugeskrift for Læger* 2018;180 V06170481.

Juslin, P. N., & Timmers, R. (2010). Expression and communication of emotion in music. In P. N. Juslin & J. Sloboda (Eds.), *Handbook of music and emotion: Theory, research, applications.* New York: Oxford University Press.

Kattenstroth, J.-C., Kalisch, T., Holt, S., Tegenthoff, M., & Dinse, H. R. (2013). Six months of dance intervention enhances postural, sensorimotor and cognitive performance in elderly without affecting cardio-respiratory functions. *Frontiers in Aging Neuroscience.* Feb 2013. https://doi.org/10.3389/fnagi.2013.00005.

Krantz, G., Madison, G., Merker, B. (2006). Melodic intervals as reflected in body movement. *Ninth International Conference on music perception and Cognition*, Bologna Aug 22–26.

Krantz, G., Kreutz, G., Ericson, M., & Theorell, T. (2010). Bodily movements influence heart rate variability (HRV) responses to isolated melodic intervals. *Music and Medicine, 3*(2), 108–113.

Kreutz, G. (2014). Does singing facilitate social bonding? *Music and Medicine, 6*, 51–60.

Kreutz, G., Bongard, S., Rohrmann, S., Hodapp, V., & Grebe, D. (2004). Effects of choir singing or listening on secretory immunoglobulin A, cortisol, and emotional state. *Journal of Behavioral Medicine, 27*, 623–635.

Krumhansl, C. L. (1997). An exploratory study of musical emotions and psychophysiology. *Canadian Journal of Experimental Psychology, 51*, 336–352.

Le Doux, J. (1998). *The emotional brain*. New York: Weidenfeld and Nicolson.

Lewis, A., Cave, P., Stern, M., Welch, L., Taylor, K., Russell, J., et al. (2016). Singing for lung health-a systematic review of the literature and consensus statement. *NPJ Primary Care Respiratory Medicine, 26*, 16080. https://doi.org/10.1038/npjpcrm.2016.80.

Lindblad, F., Hogmark, Å., & Theorell, T. (2007). Music intervention for 5th and 6th graders – Effects on development and cortisol secretion. *Stress and Health, 23*(1), 9–14.

Lingham, J., & Theorell, T. (2009). Self-selected "favourite" stimulative and sedative music listening – how does familiar and preferred music listening affect the body? *Nordic Journal of Music Therapy, 18*, 150–166.

Merriam, A. P. (1964). *The anthropology of music*. Evanston: Northwestern Univ. Press.

Miller, M., Mangano, C., Beach, V., Kop, W., & Vogel, R. A. (2010). Divergent effects of joyful and anxiety-provoking music on endothelial vasoreactivity. *Psychosomatic Medicine, 72*(4), 354–356. https://doi.org/10.1097/PSY.0b013e3181da7968. Epub 2010 Apr 5.

Moon, C., Lagercrantz, H., & Kuhl, P. K. (2013). Language experienced in utero affects vowel perception after birth: A two-country study. *Acta Paediatrica, 102*, 156–160.

Morris, J. S., Öhman, A., & Dolan, R. J. (1998). Conscious and unconscious emotional learning in the human amygdala. *Nature*., 4, *393*(6684), 467–470.

Olsson, E. M. G., von Schéele, B., & Theorell, T. (2013). Heart rate variability during choral singing. *Music and Medicine, 5*, 52–59. https://doi.org/10.1177/1943862112471399.

Pallesen, K. J., Brattico, E., Bailey, C., Korvenoja, A., Koivisto, J., Gjedde, A., et al. (2005). Emotion processing of major, minor, and dissonant chords: A functional magnetic resonance imaging study. *Annals of the New York Academy of Sciences, 1060*, 450–453.

Porges, S. W. (2007). The polyvagal perspective. *Biological Psychology, 74*, 116–143.

Quiroga, M. C., Bongard, S., & Kreutz, G. (2009). Emotional and neurohumoral responses to dancing tango Argentino: The effects of music and partner. *Music and Medicine, 1*, 14–21.

Sifneos, P. E. (1996). Alexithymia: Past and present. *American Journal of Psychiatry, 153*(suppl), 137–142.

Spychiger, M. (2002). Music education is important – why? In G. Matell & T. Theorell (Eds.), *Barn och Musik*. Stockholm: Stressforskningsrapporter, Stressforskningsinstitutet, Stockholms Universitet.

Szmedra, L., & Bacharach, D. W. (1998). Effect of music on perceived exertion, plasma lactate, norepinephrine and cardiovascular hemodynamics during treadmill running. *International Journal of Sports Medicine, 19*, 32–37.

Theorell, T. (2014). *Psychological health effects of musical experiences. Theories, studies and reflections in music health science*. Dordrecht: Springer.

Theorell, T., Konarski, K., Westerlund, H., Burell, A. M., Engström, R., Lagercrantz, A. M., et al. (1998). Treatment of patients with chronic somatic symptoms by means of art psychotherapy: A process description. *Psychotherapy and Psychosomatics, 67*, 50–56.

Ullén, F., de Manzano, Ö., Harmat, L., & Theorell, T. (2010). The physiology of effortless attention: Correlates of state flow and flow proneness. In B. Bruya (Ed.), *Effortless attention. A new perspective in the cognitive science of attention and action* (pp. 205–217). Cambridge, MA: The MIT Press.

Vickhoff, B., Åström, R., Theorell, T., & von Scheele, B. (2012). Musical piloerection. *Music and Medicine, 4*(2), 82–89.

Vickhoff, B., Malmgren, H., Åström, R., Nyberg, G., Engvall, M., Snygg, J., et al. (2013). Music determines heart rate variability of singers. *Frontiers in cognitive neuroscience*. https://doi.org/10.3389/psyg.2013.00334.

Chapter 6
How Music and Social Engagement Provides Healthy Aging and Prevents Behavioural and Psychological Symptoms of Dementia

Hanne Mette Ridder

Introduction

There is no direct link between aging and dementia, as dementia is a syndrome not caused by age itself; however, the risk of getting dementia increases with age. The way we age will, together with genetic factors, influence the development of pathological changes in the brain and also influence how the symptoms of these changes are expressed. This is explained by the theory of cognitive reserve and the understanding of how the internal brain stem regulation system empowers mental effort and social interactions in a way that challenges in a positive manner. With the release of norepinephrine, age-related brain deterioration is prevented and cognitive function maintained. Norepinephrine is a neurotransmitter released from the locus coeruleus and plays a role not only in social engagement and learning but also in the development of Alzheimer's disease. In the following chapter, I will present this theoretical perspective on sensory processing and brain stem functioning in order to understand why music plays such a remarkable role in healthy aging, in the prevention of dementia symptoms and in keeping the person with advanced dementia socially engaged.

Healthy Aging

Populations live longer around the world, regardless of living in high- or low-income countries. Longevity is a great opportunity for modern societies but only if the extra years are "worth" living for the person. Health problems in old age may be a consequence of chronic disease and lead to reduced well-being; however such

H. M. Ridder (✉)
Department of Communication and Psychology, Aalborg University, Aalborg, Denmark
e-mail: hanne@hum.aau.dk

L. O. Bonde, T. Theorell (eds.), *Music and Public Health*,
https://doi.org/10.1007/978-3-319-76240-1_6

problems may be prevented or delayed. We know that physical activity and the right diet brings health, but we still know too little about lifelong mental and cognitive health. According to the WHO (2015), healthy aging requires that the person remains active, autonomous and integrated. It also requires a "process of developing and maintaining the functional ability that enables well-being in older age" (p. 228). Functional ability is determined by environmental factors *and* the person's intrinsic capacity. Environmental factors include anything from policies, systems, services and technologies. Among these, social relationships and cultural and social values are mentioned. Intrinsic capacities are the "combination of all the individual's physical and mental capacities" (WHO 2016a, p. 4).

Putting healthy aging on the political agenda seems a burning issue, but at the same time, we must carefully consider and question how we achieve healthy aging in modern societies. On the one hand, age is often stigmatized, and we witness ageism, anti-aging ideals and gerontophobia. On the other hand, what we understand as good health is highly subjective, and a top-down regulated system that forces restrictions to everyday living may not have the expected benefits or encompass what we understand as good living.

The key actions in developing healthy aging and age-friendly environments include fostering older persons' autonomy and enabling engagement according to the WHO 5-year global strategy and action plan (WHO 2016a, p. 7). The WHO highlights "integrated and person-centred approaches" and suggests that such approaches have better outcomes without being more expensive than traditional services (WHO 2015, p. 115). Continued personal growth is key when it comes to enabling older people to do what they value and to stay autonomous. Personal growth means continuing to learn and apply knowledge, engage in problem solving, be able to make choices and keep a sense of control (p. 174). In addition, social and emotional growth is imperative for how health is perceived. According to the WHO, social and emotional growth has the chance to increase with age *if* the population has opportunities to improve self-knowledge, self-regulation skills and social relationships over the years (p. 174).

Loneliness is linked with decreases in health status and quality of life; therefore reciprocal relationships are regarded as important for the person's sense of self-worth and as a motivator for continued social engagement (WHO 2015, 188). Consequently, it becomes increasingly important to maintain relationships. If we aim to meet the societal response to healthy aging, this requires "a transformation of health systems that moves away from disease-based curative models" and moves towards person-centred and integrated care (p. 223).

Growth and Cognitive Reserve

Introducing the concept of personal growth for the aged person is in line with the developmental psychologist Erik Erikson (Erikson and Erikson 1998) who described old age as a period to achieve integrity. Integrity contributes to wisdom and is opposed to stagnation and despair. In order to overcome and/or accept despair, the person must be able to interact with others without being absorbed in own needs but concerned and interested in matters beyond these. Erikson even coined the term gerotranscendence. Transcending "gero" (old age) involves soul and body and the surpassing of all human knowledge and experience; achieving this first of all demands honesty and humility (Erikson and Erikson 1998). In a Western society, *learning* has often been understood as processes relevant only in childhood and adolescence, with adults having "achieved" learning and therefore with difficulties in learning new skills. In fact, with age learning does take longer time, but at the same time, the old person may better store new learned knowledge as it will connect to consolidated learning from earlier experiences. This is what *wisdom* is about and described as a late-emerging mental strength (Goldberg 2005).

The general view on learning changed with the breakthrough research on adult neurogenesis in the end of the twentieth century. The belief that humans are born with a restrained quantity of neurons was rejected when it was discovered that neurons are reproduced throughout life in specific brain areas such as in the hippocampus (Eriksson et al. 1998; Kuhn et al. 1996). In the same time period, it was also discovered that the degree of brain pathology observed at autopsy did not explain the degree of clinical manifestation of Alzheimer's disease (AD). The term cognitive reserve was devised to explain why some persons with AD perform perfectly well in cognitive tests despite clear neuropathological markers of AD (Stern 2009). It is assumed that those who are able to perform in spite of severe damage have a better connected brain, with mechanisms able to reorganize around the AD pathology (Robertson 2013). Four of the most common elements to improve cognitive reserve are according to the neuropsychologist and stress researcher Ian Robertson (2013) education level, mental activity, social engagement and enriched/novel environments.

Norepinephrine, Locus Coeruleus and Arousal

The elements contributing to cognitive reserve (education, mental activity, social engagement and enriched environments) all have an upregulating effect on the norepinephrine (NE) system. NE, also called noradrenaline, is a neurotransmitter and neuromodulator produced in a small area in the brain stem called the locus coeruleus. Its pathways connect to multiple brain regions such as the cerebellum, the forebrain and the spinal cord and regulate mood, memory, hormones, cerebral blood flow and motor behaviour (Zillmer and Spiers 2001). As NE is depleted with age, upregulation of NE may not only increase attention and learning but also reduce the risk of AD. NE, therefore, "is a strong candidate for mediating the compensatory aspect of cognitive reserve" (Robertson 2013, p. 304).

Dysfunction in the locus coeruleus and NE system is described to cause abnormalities in arousal level, leading to psychiatric symptoms and psychosis (Yamamoto et al. 2014), attention-deficit/hyperactivity disorder and depression (Wilson et al. 2013), and accelerate physiological aging (Williamson et al. 2015). Negative stress may lead to dysfunction; however, positive stress has the opposite effect. When NE is released during arousing, mentally challenging or novel situations that are perceived as *positively exciting*, interesting and stimulating, neurons are protected from damage. This may explain how education and engaging careers prevent or delay cognitive decline in aging and suggests that "cognitive challenges and physical exercise may be effective interventions throughout life that harness the anti-inflammatory and cell-protective qualities of NE to help to forestall cognitive decline and dementia" (Mather and Harley 2016, p. 224).

When a person experiences features of safety, this leads to autonomic reactions that promote "open receptivity with others" (Williamson et al. 2015, p. 4). A person's reactions to what is perceived as either threatening or safe are part of an evolutionary adaptive response in humans and activated prior to conscious awareness (Lanius et al. 2017; Porges 2009; Porges and Lewis 2010). This happens through a network of interconnected brain regions including the locus coeruleus. Consequently, by engaging in mental effort and social interactions in a way that challenges in a positive and safe way, the NE release will prevent age-related damage, maintain cognitive function for a longer time and open a window for social interaction.

Mental effort is according to Daniel Kahneman correlated with arousal and pupil size (Kahneman and Peavler 1969). Kahneman and his team used pupillary changes to investigate a wide range of psychological phenomena. When the pupils, the black spot in our eyes, widens (dilates) this is not only related to the brightness of what we see but also related to the difficulty or complexity of a task we are imagining or facing. As NE is regulating arousal levels, a relationship between the NE system, pupil size changes and human memory encoding is suggested (Hoffing and Aaron 2015). We may use this information to understand responses to music and how music may facilitate social interaction.

Dementia

When an older person experiences problems with remembering, comprehending, judging, thinking and learning, it has previously been regarded as a normal part of the aging process and that becoming "senile" is what all old people may expect. However, dementia is now regarded as not being a part of normal aging but caused by a range of different diseases (e.g. Alzheimer's disease) that leads to pathological changes and cell death in brain tissue. Although we know much more about dementia diseases, modern health care still struggles to provide integrated and person-centred care that allows for autonomy, engagement and personal growth. According to the WHO (2016b), persons with dementia are often denied basic rights and freedom, e.g. with physical and chemical restraints used extensively in care facilities.

Sensory Decline

The pathological changes in the brain due to dementia do not only affect how the person performs intellectually or bodily but in all aspects of performing in daily life, therefore, also in how the person senses. Sense perception is the process of "knowing" and begins with arousal and orientation and then sensation and finally perception (Zillmer and Spiers 2001). With sense perception, we organize, identify and interpret the sensory information and with the aim to understand where we are situated and what goes on in others, in ourselves and in the environment. Disturbed sensory processing makes it very challenging to understand sensory input and to make sense of situations and interactions.

Problems with sensation and perception may increase in normal aging, but is assumed to have accelerated progress in dementia. As an example, impaired auditory processing is observed to a larger degree in persons with AD than in a nonclinical population and may therefore be an early marker of AD (Daulatzai 2016). Sensory decline makes activities of daily living more challenging and may as a consequence inhibit the person in outgoing social activities. The consequences are often stress reactions or social isolation which may speed up the progression of dementia, leading to further cognitive decline. This is underlined by the fact that nursing home residents with dual sensory impairment (i.e. of concurrent vision and hearing) show greater cognitive decline over time compared to those without sensory impairment (Daulatzai 2016).

As an example, we may seek eye contact to be sure we are correctly understood or to be reassured. With visual impairment, we will miss such communicative signals, and we may more easily feel ignored or even rejected.

Behavioural and Psychological Symptoms of Dementia

To feel acknowledged is a need all human beings experience. According to dementia researcher Tom Kitwood (1997), such rudimentary needs are grounded in our evolutionary past and mediated by basic nervous system functioning. He defined a cluster of psychosocial needs (comfort, attachment, inclusion, occupation and identity) with all five needs coming together in the need for love. If the psychosocial needs are not met, the person with dementia will communicate them in other ways. With missing abilities to communicate and express oneself, the attempts to communicate needs may sometimes be difficult to understand for others. They may be misinterpreted and described as inappropriate or disturbing behaviour. It is difficult to define when such behavioural and psychological symptoms of dementia (BPSD) are caused by brain pathology or by communication related to psychosocial needs. BPSD is negatively associated with both patient and carer ratings of quality of life (Hurt et al. 2008). For future research regarding BPSD, it is recommended to focus on research on movement-based therapies, hands-on (touch) therapies and interventions provided during personal care routines and with these interventions tailored to balance individual arousal patterns (Kverno et al. 2009).

Eye Contact and Trust

Previously eye contact was described as important for feeling acknowledged, but for a person in a state of high arousal, direct eye contact may be perceived as a threat, and instead of bringing trust and resulting in meeting the psychosocial needs of the person, eye contact may lead to fight or flight responses (Lanius et al. 2017). This reminds us how important caregiver competences are. Caregivers must not only be able to understand when the person with dementia communicates psychosocial needs but also understand and accordingly react to their level of arousal. In one situation mutuality and eye contact help the person with dementia, in another, regulation and respectful distance.

Alzheimer's Disease and the Locus Coeruleus

Alzheimer's disease is the most common form of dementia. Certain patterns of protein deposition (amyloid and tau) are formed in the brain of persons with AD and progressively destroy neurons and the communication between them. These impairments have been observed to spread out from brain areas around the hippocampus. Hippocampus is a structure in the medial temporal lobe and plays a critical role in context-rich and context-free memory (Didic et al. 2011). Recent research, however, has shown that the locus coeruleus is particularly exposed to impairment and may be the first site exhibiting AD pathology (Mather and Harley 2016).

In order to understand the AD-related pathologic changes, scientists from Germany re-examined autopsy samples of tau protein from 2332 brains of persons from 1 to 100 years of age (Braak et al. 2011). They found abnormal tau already in childhood but did not see the negative influences of amyloid until later in life. All samples from individuals older than 40 years showed a pretangle stage with the changes developing from the locus coeruleus but still not yet leading to cell death. It seems such pretangle stage may last for decades, even five or more decades, with some samples from persons older than 90 years showing only a first stage of pathological neurofibrillary tangles. From this the researchers conclude that old age does not automatically lead to AD but that the pathologic development of tangles is "an uncommonly slowly progressive one that frequently extends into old age" (Braak et al. 2011, p. 967). This finding suggests that our efforts to prevent AD should maybe start already "during the first decades of life, by protecting caeruleus projection cells and/or preventing them from developing the pretangle material" (p. 968).

The locus coeruleus is described to be related to major depression in AD (Zweig et al. 1988) and aggressive behaviour (Herrmann et al. 2004). Herrmann and colleagues describe how NE is leading to sympathetic flight or fight response, and they carried out a review to examine pharmacological treatment of BPSD in AD. They found the study of the neurobiology of BPSD in AD fraught with methodological difficulties and only identified little research into the use of NE treatments for BPSD (Herrmann et al. 2004). In a recent publication, however, "the sensory hypothesis" is applied to explain how sensory decline promotes cognitive decline, leading to AD (Daulatzai 2016). Sensory stimulation of many different modalities activates the locus coeruleus and enhances NE. With this understanding as a starting point, it is relevant to examine if music, which may integrate auditory and other sensory stimulation, plays a role in healthy aging and in preventing AD and BPSD.

Music

According to the WHO (2015), continued personal growth is not only stimulated by employment or voluntary work but also by engagement in cultural and social activities. When people listen to music, it is mainly for regulating arousal and mood and for achieving self-awareness (Schäfer et al. 2013), whereas active involvement in music or musicking is described to play an important role for emotion regulation, identity formation, and engaging in cultural and social activities (DeNora 2000; Juslin and Sloboda 2010; Small 1998; Wigram et al. 2002). Several studies also point at music as a proponent for developing cognitive reserve and for preventing behavioural and psychological symptoms of dementia, which I will address in the following.

Listening to and Practicing Music

As described above, disturbed sensory processing may lead to difficulties in carrying out activities of daily living and/or to stress reactions, both leading to social isolation. In order to be able to socially engage in cultural activities, the processing of auditory stimuli plays a role. In a place with several people and lots of noise, it is challenging to focus on what you want to listen to and to ignore irrelevant sounds. Such complex listening abilities are better preserved in musicians, and musical training may therefore provide cognitive reserve (Alain et al. 2014; Fauvel et al. 2013; Seinfeld et al. 2013; Zendel and Alain 2012). This is in line with a study of 237 cognitively intact participants from whom self-reported formal musical training in early- to midlife was associated with improved late-life episodic and semantic memory performance. Intensive music training in early childhood leads to structural changes in the brain (Hyde et al. 2009) and is therefore regarded as a path to cognitive reserve (Gooding et al. 2014). The explanation could be that practicing music leads to better executive control later in life (Moussard et al. 2016).

It is complex to play music. As an example, pianists integrate the reading of music scores and movements of the fingers into an aesthetic and temporal context. This is very difficult in the beginning and demands lots of rehearsal, but with time, neural networks are stimulated and skills become implicit and can be performed with much less attention and effort (Bugos et al. 2007). In this way music may boost sensorimotor and cognitive functioning which suggests "the potential for music making as an interactive treatment or intervention for neurological and developmental disorders, as well as those associated with normal aging" (Wan and Schlaug 2010).

Music and singing facilitate learning and preservation of learned material in persons with AD, making music effective as a mnemonic technique (Palisson et al. 2015; Simmons-Stern et al. 2010). The benefits may play a role not only in the present moment but across the lifespan. Music training is described as an activity that modifies a hierarchy of brain structures, offering "distinct perceptual and cognitive benefits not observed with other forms of intense training or experience" (Moreno and Bidelman 2014, p. 94).

Dancing

Practicing a musical instrument involves movements of parts of the body in a temporal context, as does dancing. In a study by Porat et al. (2016), a group of participants with and without mild cognitive impairment (MCI) with a mean age of 70 years (ranging from 51 to 90 years) identified themselves as dancers ($n = 44$); they had engaged in either amateur dancing or formal dance training, some in their childhood, some still actively dancing. Cognitive tests and fMRI scans were compared with a group ($n = 43$) who reported having no dance experience. Although there were no significant differences between the groups in age, sex, education, or score on cognitive functioning, there was a higher proportion of persons with MCI between the nondancers. Further, the dancers performed better in learning and memory tasks and showed significant thinner cortical grey matter (Porat et al. 2016). These findings suggest that dance may provide cognitive reserve and correspond with a cohort study showing that leisure activities (reading, playing board games, playing musical instruments and dancing) reduce the risk of developing dementia (Verghese et al. 2003).

Music and Memory

Music is described as having the power to unlock memories and other cognitive capacities in AD (Clark and Warren 2015; Innes et al. 2017) and seems to involve distinct and task-dependent memory systems (Jacobsen et al. 2015). The network engaged in encoding musical memory is partly independent of other memory systems, and by comparing these brain areas with those affected by AD, researchers found the regions normally involved in musical memory encoding strikingly well preserved in AD (Jacobsen et al. 2015). Twin studies may in particular answer questions about music's cognitive reserve capacity, and in this context a study from Sweden is interesting as it compared those twins where only one played a musical instrument. After controlling for various demographic factors, the study showed that playing an instrument significantly decreases the risk of developing cognitive impairment and dementia, with a difference that seems to be due to music training and not only genetic factors (Balbag et al. 2014).

Music and Pupil Response

Pupil dilation may be an indicator of activation of the NE system and further to engagement and emotional response. Gingras et al. (2015) used an infrared optical eye-tracking system to explore music listening in nonmusicians ($n = 30$). The participants rated their level of arousal, pleasantness, tension and familiarity with the music while listening to short 6-second music excerpts. The study showed that arousal and tension ratings were significantly correlated with mean pupillary response, and that larger dilations were an indicator of music playing a greater role in life for the participant. Thus, if the level of arousal and tension in music listeners correlates with pupil dilation, the role of music in everyday living may be predicted through eye tracking, taking into account that "responses to music depend on characteristics of the listener as well as on the music itself" (Gingras et al. 2015, p. 9).

The role of listener preferences is confirmed by Laeng et al. (2016) who compared music-induced aesthetic "chill" responses with eye tracker measured pupil diameter during music listening ($n = 52$, mean age 32; range 21–59). They found that self-selected songs resulted in more chills than in songs selected by others and that pupil diameter mirrors intense responses to music (p. 172). Based on the pupillary responses, the researchers suggest a neuromodulatory role of the central NE system. This leads to the assumption that engagement in music not only is about the aesthetics of the music in itself but about the listener's relation to the music as well as individual preferences. Through their research in music chills, Laeng and colleagues describe the pupil as a mirror to music's soul, with the specific characteristics of music affecting listeners differently.

In summing up from this first part of the chapter, we may claim that engagement via positively challenging mental effort prevent age-related cognitive decline. This may explain why preferred music through the release of NE mediates healthy aging. With the upregulation of NE, music has a compensatory effect and may play a role in building up cognitive reserve and preventing symptoms of dementia.

Music and Social Engagement for Preventing BPSD

Among many other functions, music is described to play an important role for social engagement in older people or in dementia (Creech et al. 2013; Hallam et al. 2014; McDermott et al. 2014; Ridder 2011; Spiro 2010). Further, apart from having a preventive effect on cognition and dementia, music may reduce behavioural and psychological symptoms of dementia (Guzmán-García et al. 2013; Ridder et al. 2013; Särkämö et al. 2012; Sakamoto et al. 2013). In a meta-review of 34 music therapy studies, a significant effect of music therapy on disruptive behaviour and anxiety was found, as well as a positive trend for cognitive function (Zhang et al. 2017). An effect of music-based therapeutic interventions on BPSD was not confirmed in a recent Cochrane review, but an effect was confirmed for depression (van der Steen et al. 2017). Further, it was explained that "music therapists are specially qualified to work with individuals or groups of people, using music to try to help meet their physical, psychological and social needs" (p. 2).

As mentioned above music affects listeners differently according to personal preferences, and I therefore want to underline that there is no direct cause and effect relationship between music and reduction of BPSD. What I suggest here is a theoretical understanding of how social engagement, facilitated through music experiences, will allow for connecting with the person which enables meeting the psychosocial needs of the person. When psychosocial needs are met, those behavioural and psychological symptoms – understood as ways of communicating these needs – will diminish, leading to a reduction in ratings of BPSD.

For a person with dementia, the progressive loss of cognitive functions and disturbances of sensory processing impair the ability to understand and make sense of a situation and the intentions of a caregiver or others. For some persons with dementia, it becomes difficult to self-regulate and to feel acknowledged as a person. At an overall level, it becomes extremely challenging to remain active, autonomous and integrated. This affects quality of life and how health is perceived. For many, music could be a key to engage socially. This may be music experiences in the form of singing, dancing or playing instruments or expression through sounds. Or it may be through music listening by sharing emotional expressions inspired or stimulated from the music.

In a review of literature on human communication, Kraus and Slater (2016) explain how our modern languages and musical systems reflect human development in a world of sound (p. 84). Communication is a product of human cognitive abilities to make sense of sound and movements in order to construct a representation of what happens around us. Neurons adjust their firing to not only language but also to musical systems, making human brains innately wired with complex pattern detection mechanisms (p. 87). These mechanisms are present in newborns and seem to be preserved in dementia, although differently. Some persons with severe Alzheimer's disease show fully normal musical abilities, whereas others show only partial preservation of the ability to perform musical tasks. For some, these abilities are lost (Vanstone and Cuddy 2010). Still, musical memory is surprisingly well preserved in many persons with AD (Jacobsen et al. 2015). For those individuals where the ability to enjoy and respond to music is still present, "this preservation could serve as an important avenue to enhanced quality of life for a group of people who have lost so many other abilities" (Vanstone and Cuddy 2010, p. 125). What may remain is the ability to tap into the inherent rhythms of music which may facilitate interpersonal synchrony (p. 97). This allows the person with dementia to predict what is coming next and to build social bonds. Such interpersonal synchrony is what communication is all about: "Words may scratch the surface, but sound can move us beyond words" (Kraus and Slater 2016, p. 97).

Conclusion

If interpersonal synchrony and the communicative sharing it fosters are meaningful to the person with dementia, we may expect the person to be mentally challenged in a positive way. The ability to relate to music is often preserved in persons with dementia, and this ability hereby provides an opportunity for social engagement. We may further expect this to affect brain stem systems, more specifically the locus coeruleus, and lead to the release of NE. Even if the person is not able to communicate verbally about the music experience or seems to have forgotten it shortly after due to severe dementia, this may explain why the confident and safe social interaction has a stabilizing function and the potential to positively affect overall health and well-being of the person and potentially lead to a reduction of BPSD. This understanding may help as an argument for promoting psychosocial care interventions integrating music activities, musical communication methods and music therapy. In addition, it may push our understanding of learning, calling for actions to integrate music and other engaging activities in our lives from early childhood. The preventive mechanisms from the positive influence on social engagement, education, mental activity and enriched environments contribute to promote good living, also in old age.

References

Alain, C., Zendel, B. R., Hutka, S., & Bidelman, G. M. (2014). Turning down the noise: The benefit of musical training on the aging auditory brain. *Hearing Research, 308*, 162–173. https://doi.org/10.1016/j.heares.2013.06.008.

Balbag, M. A., Pedersen, N. L., & Gatz, M. (2014). Playing a musical instrument as a protective factor against dementia and cognitive impairment: A population-based twin study. *International Journal of Alzheimer's Disease, 2014*, 1–7. https://doi.org/10.1155/2014/836748.

Braak, H., Thal, D. R., Ghebremedhin, E., & Del Tredici, K. (2011). Stages of the pathologic process in Alzheimer disease: Age categories from 1 to 100 years. *Journal of Neuropathology & Experimental Neurology, 70*(11), 960–969. https://doi.org/10.1097/NEN.0b013e318232a379.

Bugos, J. A., Perlstein, W. M., McCrae, C. S., Brophy, T. S., & Bedenbaugh, P. H. (2007). Individualized piano instruction enhances executive functioning and working memory in older adults. *Aging & Mental Health, 11*(4), 464–471. https://doi.org/10.1080/13607860601086504.

Clark, C. N., & Warren, J. D. (2015). Music, memory and mechanisms in Alzheimer's disease. *Brain, 138*(8), 2122–2125.

Creech, A., Hallam, S., Varvarigou, M., McQueen, H., & Gaunt, H. (2013). Active music making: A route to enhanced subjective well-being among older people. *Perspectives in Public Health, 133*(1), 36–43. https://doi.org/10.1177/1757913912466950.

Daulatzai, M. A. (2016). Dysfunctional sensory modalities, locus coeruleus, and basal forebrain: Early determinants that promote neuropathogenesis of cognitive and memory decline

and Alzheimer's disease. *Neurotoxicity Research, 30*(3), 295–337. https://doi.org/10.1007/s12640-016-9643-3.
DeNora, T. (2000). *Music in everyday life*. Oxford: Oxford University Press.
Didic, M., Barbeau, E. J., Felician, O., Tramoni, E., Guedj, E., Poncet, M., & Ceccaldi, M. (2011). Which memory system is impaired first in Alzheimer's disease? *Journal of Alzheimer's Disease, 27*(1), 11–22. https://doi.org/10.3233/JAD-2011-110557.
Erikson, E. H., & Erikson, J. M. (1998). *The life cycle completed (extended version)*. New York: WW Norton & Company.
Eriksson, P. S., Perfilieva, E., Bjork-Eriksson, T., Alborn, A. M., Nordborg, C., Peterson, D. A., et al. (1998). Neurogenesis in the adult human hippocampus. *Nature Medicine, 4*(11), 1313–1317. https://doi.org/10.1038/3305.
Fauvel, B., Groussard, M., Eustache, F., Desgranges, B., & Platel, H. (2013). Neural implementation of musical expertise and cognitive transfers: Could they be promising in the framework of normal cognitive aging? *Frontiers in Human Neuroscience, 7*, 693. https://doi.org/10.3389/fnhum.2013.00693.
Gingras, B., Marin, M. M., Puig-Waldmüller, E., & Fitch, W. T. (2015). The eye is listening: Music-induced arousal and individual differences predict pupillary responses. *Frontiers in Human Neuroscience, 9*, 619. https://doi.org/10.3389/fnhum.2015.00619.
Goldberg, E. (2005). *The wisdom paradox: How your mind can grow stronger as your brain grows older. Library journal*. London: The Free Press.
Gooding, L. F., Abner, E. L., Jicha, G. A., Kryscio, R. J., & Schmitt, F. A. (2014). Musical training and late-life cognition. *American Journal of Alzheimer's Disease and Other Dementias, 29*(4), 333–343. https://doi.org/10.1177/1533317513517048.
Guzmán-García, A., Hughes, J. C., James, I. A., & Rochester, L. (2013). Dancing as a psychosocial intervention in care homes: A systematic review of the literature. *International Journal of Geriatric Psychiatry, 28*(9), 914–924. https://doi.org/10.1002/gps.3913.
Hallam, S., Creech, A., Varvarigou, M., McQueen, H., & Gaunt, H. (2014). Does active engagement in community music support the well-being of older people? *Arts & Health, 6*(2), 101–116. https://doi.org/10.1080/17533015.2013.809369.
Herrmann, N., Lanctôt, K. L., & Khan, L. R. (2004). The role of norepinephrine in the behavioral and psychological symptoms of dementia. *The Journal of Neuropsychiatry and Clinical Neurosciences, 16*(3), 261–276. https://doi.org/10.1176/appi.neuropsych.16.3.261.
Hoffing, R. C., & Aaron, R. S. (2015). Pupillometry as a glimpse into the neurochemical basis of human memory encoding Russell. *Journal of Cognitive Neuroscience*. 2, *27*(4), 765–774.
Hurt, C., Bhattacharyya, S., Burns, A., Camus, V., Liperoti, R., Marriott, A., et al. (2008). Patient and caregiver perspectives of quality of life in dementia: An investigation of the relationship to behavioural and psychological symptoms in dementia. *Dementia and Geriatric Cognitive Disorders, 26*(2), 138–146. https://doi.org/10.1159/000149584.
Hyde, K. L., Lerch, J., Norton, A., Forgeard, M., Winner, E., Evans, A. C., et al. (2009). Musical training shapes structural brain development. *The Journal of Neurosicence, 29*(10), 3019–3025. https://doi.org/10.1523/JNEUROSCI.5118-08.2009.
Innes, K. E., Selfe, T. K., Khalsa, D. S., & Kandati, S. (2017). Meditation and music improve memory and cognitive function in adults with subjective cognitive decline: A pilot randomized controlled trial. *Journal of Alzheimer's Disease, 56*, 899–916. https://doi.org/10.3233/JAD-160867.
Jacobsen, J. H., Stelzer, J., Fritz, T. H., Chételat, G., La Joie, R., & Turner, R. (2015). Why musical memory can be preserved in advanced Alzheimer's disease. *Brain, 138*(8), 2438–2450. https://doi.org/10.1093/brain/awv135.
Juslin, P. N., & Sloboda, J. A. (Eds.). (2010). *Handbook of music and emotion: Theory, research, applications*. Oxford: Oxford University Press.
Kahneman, D., & Peavler, W. S. (1969). Incentive effects and pupillary changes in association learning. *Journal of Experimental Psychology, 79*(2), 312–318. https://doi.org/10.1037/h0026912.

Kitwood, T. (1997). *Dementia reconsidered. The person comes first*. Buckingham: Open University Press.

Kraus, N., & Slater, J. (2016). Beyond words: How humans communicate through sound. *Annual Review of Psychology, 67*(1), 83–103. https://doi.org/10.1146/annurev-psych-122414-033318.

Kuhn, H. G., Dickinson-Anson, H., & Gage, F. H. (1996). Neurogenesis in the dentate gyrus of the adult rat: Age-related decrease of neuronal progenitor proliferation. *Journal of Neuroscience, 16*(6), 2027–2033.

Kverno, K. S., Black, B. S., Nolan, M. T., & Rabins, P. V. (2009). Research on treating neuropsychiatric symptoms of advanced dementia with non-pharmacological strategies, 1998–2008: A systematic literature review. *International Psychogeriatrics/IPA, 21*(5), 825–843. https://doi.org/10.1017/S1041610209990196.

Laeng, B., Eidet, L. M., Sulutvedt, U., & Panksepp, J. (2016). Music chills: The eye pupil as a mirror to music's soul. *Consciousness and Cognition, 44*, 161–178. https://doi.org/10.1016/j.concog.2016.07.009.

Lanius, R. A., Rabellino, D., Boyd, J. E., Harricharan, S., Frewen, P. A., & McKinnon, M. C. (2017). The innate alarm system in PTSD: Conscious and subconscious processing of threat. *Current Opinion in Psychology, 14*, 109–115. https://doi.org/10.1016/j.copsyc.2016.11.006.

Mather, M., & Harley, C. W. (2016). The locus coeruleus: Essential for maintaining cognitive function and the aging brain. *Trends in Cognitive Sciences, 20*(3), 214–226. https://doi.org/10.1016/j.tics.2016.01.001.

McDermott, O., Orrell, M., & Ridder, H. M. (2014). The importance of music for people with dementia: The perspectives of people with dementia, family carers, staff and music therapists. *Aging & Mental Health, 18*(6), 706–716. https://doi.org/10.1080/13607863.2013.875124.

Moreno, S., & Bidelman, G. M. (2014). Examining neural plasticity and cognitive benefit through the unique lens of musical training. *Hearing Research, 308*, 84–97. https://doi.org/10.1016/j.heares.2013.09.012.

Moussard, A., Bermudez, P., Alain, C., Tays, W., & Moreno, S. (2016). Life-long music practice and executive control in older adults: An event-related potential study. *Brain Research, 1642*, 146–153. https://doi.org/10.1016/j.brainres.2016.03.028.

Palisson, J., Roussel-Baclet, C., Maillet, D., Belin, C., Ankri, J., & Narme, P. (2015). Music enhances verbal episodic memory in Alzheimer's disease. *Journal of Clinical and Experimental Neuropsychology, 3395*, 1–15. https://doi.org/10.1080/13803395.2015.1026802.

Porat, S., Goukasian, N., Hwang, K. S., Zanto, T., Do, T., Pierce, J., et al. (2016). Dance experience and associations with cortical gray matter thickness in the aging population. *Dementia and Geriatric Cognitive Disorders Extra, 94704*, 508–517. https://doi.org/10.1159/000449130.

Porges, S. W. (2009). The polyvagal theory: New insights into adaptive reactions of the autonomic nervous system. *Cleveland Clinic Journal of Medicine, 76*(Suppl2), S86–S90. https://doi.org/10.3949/ccjm.76.s2.17.

Porges, S. W., & Lewis, G. F. (2010). The polyvagal hypothesis: Common mechanisms mediating autonomic regulation, vocalizations and listening. *Handbook of Mammalian Vocalization: An Integrative Neuroscience Approach, 19*, 255–264. https://doi.org/10.1016/B978-0-12-374593-4.00025-5.

Ridder, H. M. (2011). How can singing in music therapy influence social engagement for people with dementia. Insights from the polyvagal theory. In F. A. Baker & S. Uhlig (Eds.), *Voicework in music therapy. Research and practice* (pp. 130–146). London: Jessica Kingsley Publishers.

Ridder, H. M., Stige, B., Qvale, L. G., & Gold, C. (2013). Individual music therapy for agitation in dementia: An exploratory randomized controlled trial. *Aging & Mental Health, 17*(6), 667–678. https://doi.org/10.1080/13607863.2013.790926.

Robertson, I. H. (2013). A noradrenergic theory of cognitive reserve: Implications for Alzheimer's disease. *Neurobiology of Aging, 34*(1), 298–308. https://doi.org/10.1016/j.neurobiolaging.2012.05.019.

Sakamoto, M., Ando, H., & Tsutou, A. (2013). Comparing the effects of different individualized music interventions for elderly individuals with severe dementia. *International Psychogeriatrics/IPA, 25*(5), 775–784. https://doi.org/10.1017/S1041610212002256.

Särkämö, T., Laitinen, S., Tervaniemi, M., Numminen, A., Kurki, M., & Rantanen, P. (2012). Music, emotion, and dementia: Insight from neuroscientific and clinical research. *Music and Medicine, 4*(3), 153–162. https://doi.org/10.1177/1943862112445323.

Schäfer, T., Sedlmeier, P., Städtler, C., & Huron, D. (2013). The psychological functions of music listening. *Frontiers in Psychology, 4*, 511, 1–33. https://doi.org/10.3389/fpsyg.2013.00511.

Seinfeld, S., Figueroa, H., Ortiz-Gil, J., & Sanchez-Vives, M. V. (2013). Effects of music learning and piano practice on cognitive function, mood and quality of life in older adults. *Frontiers in Psychology, 4*, 810, 1–13. https://doi.org/10.3389/fpsyg.2013.00810.

Simmons-Stern, N. R., Budson, A. E., & Ally, B. A. (2010). Music as a memory enhancer in patients with Alzheimer's disease. *Neuropsychologia, 48*(10), 3164–3167. https://doi.org/10.1016/j.neuropsychologia.2010.04.033.Music.

Small, C. (1998). *Musicking: The meanings of performing and listening*. Middletown: Wesleyan University Press.

Spiro, N. (2010). Music and dementia: Observing effects and searching for underlying theories. *Aging & Mental Health, 14*(8), 891–899. https://doi.org/10.1080/13607863.2010.519328.

Stern, Y. (2009). Cognitive reserve. *Neuropsychologia, 47*(10), 2015–2028. https://doi.org/10.1016/j.neuropsychologia.2009.03.004.

van der Steen, J. T., van Soest-Poortvliet, M. C., van der Wouden, J. C., Bruinsma, M. S., Scholten, R. J. P. M., Vink, A. C. (2017). Music-based therapeutic interventions for people with dementia. *Cochrane Database of Systematic Reviews, 5*. No.: CD003477. https://doi.org/10.1002/14651858.CD003477.pub3.

Vanstone, A. D., & Cuddy, L. L. (2010). Musical memory in Alzheimer disease. *Neuropsychology, Development, and Cognition. Section B, Aging, Neuropsychology and Cognition, 17*(1), 108–128. https://doi.org/10.1080/13825580903042676.

Verghese, J., Lipton, R. B., Katz, M. J., Hall, C. B., Derby, C. A., Kuslansky, G., et al. (2003). Leisure activities and the risk of dementia in the elderly. *The New England Journal of Medicine, 348*, 2508–2516.

Wan, C. Y., & Schlaug, G. (2010). Music making as a tool for promoting brain plasticity across the life span. *The Neuroscientist: A Review Journal Bringing Neurobiology, Neurology and Psychiatry, 16*(5), 566–577. https://doi.org/10.1177/1073858410377805.

WHO. (2015). *World Health Organization. World report on ageing and health*. Geneva: World Health Organization. Retrieved from http://apps.who.int/iris/bitstream/10665/186463/1/9789240694811_eng.pdf?ua=1.

WHO. (2016a). Global strategy and action plan on ageing and health (2016–2020). A framework for coordinated global action by the World Health Organization. Geneva: World Health Organization. Retrieved from http://www.who.int/ageing/GSAP-Summary-EN.pdf?ua=1.

WHO. (2016b). *World Health Organization. Fact Sheet*. Geneva: World Health Organization. Retrieved from http://www.who.int/mediacentre/factsheets/fs362/en/.

Wigram, T., Pedersen, I. N., & Bonde, L. O. (2002). *Comprehensive guide to music therapy*. London: Jessica Kingsley Publishers Ltd..

Williamson, J. B., Porges, E. C., Lamb, D. G., Porges, S. W. (2015). Maladaptive autonomic regulation in PTSD accelerates physiological aging. *Frontiers in Psychology, 6*, 1571, 1–12. doi:https://doi.org/10.3389/fpsyg.2015.00571.

Wilson, R. S., Nag, S., Boyle, P. A., Hizel, L. P., Yu, L., Buchman, A. S., et al. (2013). Neural reserve, neuronal density in the locus ceruleus, and cognitive decline. *Neurology, 80*(13), 1202–1208. https://doi.org/10.1212/WNL.0b013e3182897103.

Yamamoto, K. I., Shinba, T., & Yoshii, M. (2014). Psychiatric symptoms of noradrenergic dysfunction: A pathophysiological view. *Psychiatry and Clinical Neurosciences, 68*(1), 1–20. https://doi.org/10.1111/pcn.12126.

Zendel, B. R., & Alain, C. (2012). Musicians experience less age-related decline in central auditory processing. *Psychology and Aging, 27*(2), 410–417. https://doi.org/10.1037/a0024816.

Zhang, Y., Cai, J., An, L., Hui, F., Ren, T., Ma, H., et al. (2017). Does music therapy enhance behavioral and cognitive function in elderly dementia patients? A systematic review and meta-analysis. *Ageing Research Reviews, 35*, 1–11. https://doi.org/10.1016/j.arr.2016.12.003.

Zillmer, E. A., & Spiers, M. V. (2001). *Principles of neuropsychology*. Belmont: Wadsworth. Thomson Learning.

Zweig, R. M., Ross, Y. I. T. C. A., Hedreen, S. J. C., Steele, C., Cardillo, M. J. E., Whitehouse, P. J., et al. (1988). The neuropathology of aminergic nuclei in Alzheimer's disease. *Annals of Neurology, 24*(2), 233–242.

Chapter 7
Music as a Forum for Social-Emotional Health

Suvi Saarikallio and Margarida Baltazar

Social-Emotional Competence (SEC) as a Link Between Music and Mental Health

Good social-emotional skills are vital throughout the life span (Charles and Carstensen 2010), and they can predict mental health and adaptation to society (Jones et al. 2015). The importance of the development of social-emotional competence starts to be noticeable already during early childhood, when children acquire skills that will be crucial for their good integration at school through healthy relationships with peers and adaptive management of affective states (Eisenberg et al. 2006; Hughes et al. 2000; Keane and Calkins 2004). The development of these skills is especially relevant during adolescence, when externalization and internalization symptomatology strongly correlates to poor social-emotional skills (Garnefski et al. 2005; Pascual et al. 2016; Seiffge-Krenke 2000; Silk et al. 2003), thus making this a crucial period for promoting the social and emotional development (Clarke et al. 2015; Steinberg 2005). The social and emotional skills are a key asset for the positive development of young people, because they increase resilience and protect against the impact of the psychosocial and environmental risk factors, thus preventing developmental trajectories that lead to behavioral and mental health problems. Indeed, the World Health Organization's (WHO) summary report on promoting mental health (2004) particularly emphasized the need for effective interventions that create psychological resources ranging from positive affect to resilience and capacity to cope with adversity.

The widely acknowledged relevance of social and emotional skills for health and well-being (e.g., Gross and John 2003) creates grounds for considering social-emotional competence as a useful concept in explaining how music can function as

S. Saarikallio (✉) · M. Baltazar
Department of Music, Art and Culture Studies, Center for Interdisciplinary Music Research, University of Jyväskylä, Jyväskylä, Finland
e-mail: suvi.saarikallio@jyu.fi

L. O. Bonde, T. Theorell (eds.), *Music and Public Health*,
https://doi.org/10.1007/978-3-319-76240-1_7

a forum to promote public health. Recent research on music's health impact has been showing that there are clear links between the frequency of music use (Thomson et al. 2014), the emotional patterns within the musical behaviors (Garrido and Schubert 2013; Thoma et al. 2012), the emotional reactivity to music (Saarikallio et al. 2013), and health outcomes. In the following sections of this chapter, we will go through some characteristics of music and the emotional and social skills linked to well-being. In particular, we discuss how music can serve as a forum for fostering the social and emotional skills and argue that this advancement of the social-emotional competence serves an *underlying mechanism* that helps to promote mental health and well-being.

The word *forum* in this context encompasses music's spatial aspect—the creation and shaping of musical environments (Born 2013). In fact, the concept of *forum* goes beyond the ancient Roman architectonical element and represents the real and virtual spaces created and delimitated by music (Clarke 2013), the inner and outer spaces it connects (Lincoln 2005; Sloboda and Juslin 2010), and the means provided by music for individual and collective expression and for the expansion of our motor, emotional, and regulative resources (Krueger 2014).

Music Is Fundamentally Social and Emotional

Music can be considered a powerful tool to raise social-emotional competence given its intrinsic connections to social and emotional experiences. The connection of music to emotion has been widely studied (Juslin &Sloboda 2001), and music is known to *express* emotions (Gabrielsson and Lindström 2010; Juslin and Laukka 2004; Juslin and Timmers 2010), which are—at least at the level of basic emotions—fast and accurately identified by listeners (Bigand et al. 2006; Juslin 1997; Juslin and Laukka 2004; Juslin and Timmers 2010; Vieillard et al. 2008). Music further *induces* a range of emotions, and the efficacy of music in evoking *positive emotions* has been indexed by self-reports (Juslin and Laukka 2004; Zentner, Grandjean & Scherer 2008); brain activation in areas connected to motivation, pleasure, and reward, including ventral striatum, dorsomedial midbrain, amygdala, hippocampus, insula, cingulate cortex, and orbitofrontal cortex (Blood and Zatorre 2001; Evers and Suhr 2000; Koelsch 2014; Menon and Levitin 2005); changes in hemispheric balance of action (Altenmuller et al. 2002; Field et al. 1998; Hausmann et al. 2016; Schmidt and Trainor 2001); and a variety of other physiological indicators: heart rate, respiration, blood pressure, skin conductivity, and biochemical responses (Bartlett 1996). In daily life, music is actively used for emotional *self-regulation* (e.g., Baltazar and Saarikallio 2016; DeNora 1999; Juslin and Laukka 2004; Saarikallio and Erkkilä 2007; Sloboda and O'Neill 2001; Thayer et al. 1994; Van Goethem 2010), and the impacts of music on reducing stress and anxiety have particularly been evidenced through effects on the major neuroendocrine system regulating these states—the hypothalamic-pituitary-adrenocortical axis (HPA). Decreases of cortisol levels, the stress hormone, have been systematically observed in relation to music listening tasks (Thoma et al. 2013), singing, playing, and dancing (Koelsch and Stegemann 2012; Kreutz et al. 2012) and music interventions (Koelsch et al. 2016; Lee et al. 2017). These findings appear observable regardless of age: the cortisol-decreasing effects, for instance, have been observed in pediatric patients and even in preterm infants in neonatal intensive care (Mörelius et al. 2016).

The use of music for *social bonding* is also a well-evidenced, universal reason for music engagement (Clayton 2009; Cross 2008, 2014; Dissanayake 2008; Mithen 2005). Infant's early interaction resembles musical expression, which supports *emotional communication* and fosters a *secure attachment* (Malloch 1999; Tafuri and Villa 2002; Trevarthen 2002; Trehub et al. 1993; Oldfield 1993; Trainor and Cirelli 2015). In youth, music listening particularly serves as a forum for peer *group identification* (Arnett 1995; North et al. 2000; Laiho 2004) and for *friendship formation* and stability (Selfhout et al. 2009). Music continues to be meaningful due to social reasons also in old age (Gembris 2008; Hays and Minichiello 2005; Laukka 2006) and strongly reflects our *belonging* to people, time, and place (Ruud 1997). Music engagement has been shown to relate to both self-reported increases in experiences of social communication and bonding (Clift and Morrison 2011; Davidson and Emberly 2012; Murcia and Kreutz 2012) and to systematic increases of hormones linked to sociality and bonding—oxytocin and β-endorphin (Kreutz et al. 2012).

What Is Social-Emotional Competence?

The concept of social-emotional competence emerges from the study of emotional intelligence and social intelligence—the types of intelligence that closely relate to the mental health and well-being outcomes (Gross and John 2003). As Bar-On (2006) points out, the development of the concepts of social and emotional intelligence show considerable overlap in terms of the involved components. Much of the emotional intelligence theories include social skills (Saarni 1990), and much of the social intelligence theories include emotional skills (e.g., Rose-Krasnor 1997).

Through these two approaches (social and/or emotional) to non-cognitive competencies, a variety of models have been formulated in parallel. For instance, Mayer and Salovey's (1997) model of emotional intelligence consists of four components: *perceiving emotions* (recognizing emotional information in faces and pictures), *understanding emotions* (knowing how emotions relate to each other and follow each other), *managing emotions* (regulation of emotions in self and in others), and *using emotions to facilitate thought* (generate a mood in the service of cognitive tasks). Bar-On's (1997) conceptualization of emotional-social intelligence also consists of four components but involves social aspects to a greater degree, consisting of the following areas: *intrapersonal* (emotional self-awareness, assertiveness, self-regard, self-actualization, independence), *interpersonal* (empathy, relationship skills, social responsibility), *adaptability* (problem-solving, reality testing, and flexibility), and *stress management* (stress tolerance, impulse control). Meanwhile, Saarni (1999) uses the term emotional competence to describe a set of skills that are relevant for child development: *awareness* of one's own emotions, ability to *discern and understand* other's emotions, ability to use the *vocabulary* of emotion and expression, capacity for *empathic* involvement, ability to *differentiate subjective emotional experience from external emotion expression*, adaptive *coping* with aversive emotions and distressing circumstances, *awareness of emotional communication* within relationships, and capacity for *emotional self-efficacy*. Further toward social competencies, Denham et al. (2012) recently integrated Rose-Krasnor's (1997) model of social competence and Payton et al.'s (2000) model of social-emotional learning to show how the specific social and emotional skills relate to each other to form the overall social-emotional competence. This model lists three emotional competence skills, *self-awareness* (identifying emotions, prosocial responsibility), *self-regulation* (managing emotions, cognition, and behavior), and *social awareness* (perspective taking, understanding emotions, caring for others), and two relational/prosocial skills, *responsible decision-making* (analyzing situations, goal setting, problem-solving) and *relationship skills* (cooperating, listening, taking turns, seeking help). According to Denham et al.'s model (2012), the above-mentioned skills are basic level fundaments that further contribute to overall success in self-regulation, social interaction, and group involvement. Indeed, many models of SEC share the idea that some of the related skills are more rudimental and basic (such as recognizing an emotion) and further serve as a foundation and building blocks for the higher-level behaviors (self-regulation, use of emotion to guide action, social interaction).

The existing models differ in terms of how closely they separate or integrate the emotional and social skills, and they also differ in whether they talk about traits or skills. The complexity of the various social, emotional, and cognitive skills involved makes SEC a highly multifaceted phenomenon. Different conceptualizations in different studies make it hard to compare and synthesize research findings. Overall, however, the existing models share the basic idea that the social-emotional competence is vital for adaptive behavior, and this makes SEC a highly relevant concept in studying health promotion.

Music and Social-Emotional Competence

Despite music's widely acknowledged connections to social and emotional *experiences*, the evidence base linking music engagement to general social-emotional *competence* is still relatively sparse. Nevertheless, preliminary studies have demonstrated some support. For instance, Theorell et al. (2014) found that musical training was related to lower alexithymia, i.e., inability to identify and describe emotions (e.g., Sifneos 1996; see Chap. 4 by Theorell and Ullén and Chap. 5 by Theorell). Musical training has also been shown to relate to higher scores in emotional intelligence (Petrides et al. 2006) and better ability in recognizing emotions in spoken sentences (Thompson et al. 2004). Accuracy of recognizing the intended emotional expression in music relates to overall empathy, particularly cognitive empathy

(Wöllner 2012) and perspective taking (Saarikallio et al. 2014). Musical training has been evidenced to make 6-year-olds better than their peers in a control group to identify difficult emotional expressions (anger and fear) in speech prosody (Thompson et al. 2004), and engagement in joint musical activity has been shown to encourage prosocial cooperative behavior in 4-year-olds (Kirschner and Tomasello 2010).

The strongest link between music and social-emotional competence in research so far has been through emotion regulation. Studies on music engagement for emotion regulation purposes show that certain uses of music reflect general emotion regulation skills and relate differently to well-being and mental health depending on the underlying strategies. The moderating role of musical strategies between music listening and its impact on affective health has been supported by self-reports (Chin and Rickard 2014; Randall et al. 2014; Thomson et al. 2014; Van den Tol and Edwards 2014) and recently also by neural responses (Carlson et al. 2015). In general, the interrelations between the used strategies and the observed symptomatology points to favoring the use of music for self-reflection, reappraisal, and distraction, in contrast to avoidant and ruminative emotion regulation strategies (Saarikallio 2017).

Overall, however, research on these connections is only emerging, and one of the major challenges in the field is that the current body of studies targets different theoretical concepts, mechanisms, experiential processes, and skills, which are not necessarily comparable or clearly defined (Baltazar and Saarikallio 2016). Social-emotional competence could thus serve as a useful overall theoretical framework to clarify the conceptual divergence in the field.

Taking into account the evidence collected on the social and emotional uses of music and their interrelations with mental health, the concept of social-emotional competence is worth exploring in the musical context. In fact, the concept of general social-emotional competence might provide a useful tool for a systematic analysis of music's impact on mental health. However, since music is a specific form of expression and experience, the theoretical frames drawn from general psychology may not fully capture the essence of what music in particular is able to offer.

Therefore, we argue for the development of a much-needed music-specific model of social-emotional competence. Evidently, in order to provide grounded hypotheses and predictions about which components of general social-emotional competence can efficiently be promoted through music, there has to be prior understanding of the key strengths of music as a forum for learning the various subcomponents of SEC. Recently, Saarikallio (2017) proposed a conceptualization for emotional health-fostering musical identity, consisting of the following elements: *emotion recognition* and identification in music, *regulation of stress and negative emotion* through music, *inducing positive emotions* and pleasure through music, *self-reflective awareness* of personal emotional responses to and use of music, and a sense of *self-agency* regarding one's emotional responses to and use of music. This model is a good example of an effort to create a music-based framework that builds on prior research on the specific nature of music as a form of emotional expression, experience, and interaction. A reprint of the model is presented in Fig. 7.1.

By analyzing music engagement through the concept of social-emotional skills, it is possible to realize that music can be a protective or a risk factor depending on

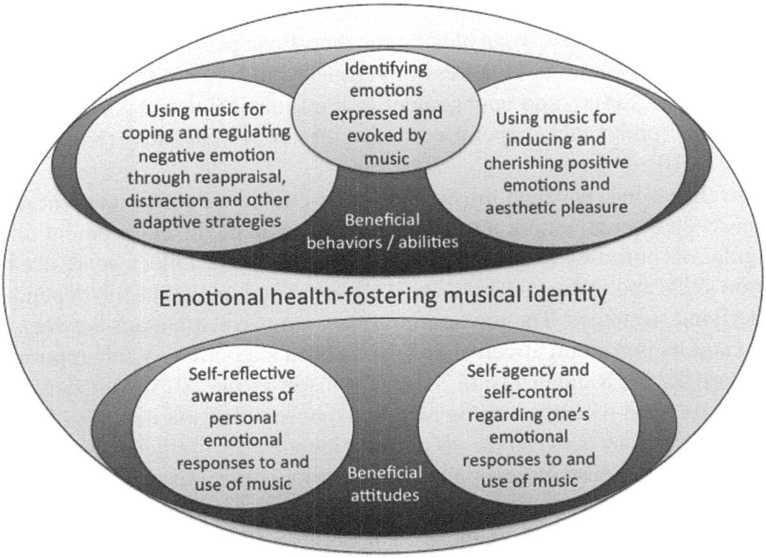

Fig. 7.1 Conceptual model of the health-relevant competencies in music. (Saarikallio 2017; reprinted with permission from the *Handbook of Musical identities*, 2017)

the competencies involved (McFerran 2016; Miranda 2013; Saarikallio 2017). A protective function of music is expected when the individual, through his/her self-agency, is able to identify the emotions expressed and induced by music, regulates the negative states mainly through self-reflection and induction of positive states, and is aware of his/her own reactions to music. Conversely, the negligent uses of music, characterized by enhancement of negative states, lack of self-reflection and awareness, and repetitive or inflexible use of regulation strategies, are a possible sign of an endangering engagement in music.

Overall, evidence is emerging that not only music engagement overall but certain music engagement patterns in particular are related to general SEC and mental health and well-being outcomes. The concept of competence fits well with the general idea that by carefully studying the impacts of particular types of music use, we can develop awareness of how music (which musical activity, for whom, in which context) most optimally serves as the forum for training general SEC and promote well-being.

Music-Based Programs for Promoting Social-Emotional Competence

Overall, the evidence base about the positive social and emotional health impacts of music, including physiological measures and rigorously conducted randomized controlled trials, is growing (MacDonald et al. 2017), and this creates ground for interventions, not only for clinical treatment but also for population-oriented primary prevention. Many general social-emotional competence promotion programs today utilize music and art engagement as their ingredients. Such programs for children, for instance, include *Papilio*, a program for primary prevention of behavioral problems and support of social-emotional competence; PATHS, Promoting Alternative Thinking Strategies program for social-emotional learning; *Incredible Years* curriculum for children's social-emotional competence; and *REDI* for school readiness. However, programs that would specifically focus on music and fully utilize the findings from current music research are rare, with only few pioneering examples. Examples of models on emotionally or socially competent versus incompetent use of music are the *model for musical emotion dysregulation* (Marik and Stegemann 2016) and the *Healthy-Unhealthy Music Scale*—HUMS (Saarikallio et al. 2015). These models provide theoretical grounds for understanding how and why music functions as a forum for training social-emotional competence and how it can be used as a tool for intervention purposes. *HUMS*, for instance, has been used to develop and test a preventive music therapy intervention for young people at risk, for improving their social, emotional, and mental well-being (Gold, Crooke, Saarikallio & McFerran 2017). *Tuned In* is another intervention program for the young, aiming for increased emotion awareness, identification, and regulation through group work on music listening patterns (Dingle & Fay 2016).

Confirming the need to further develop programs for the fostering of social-emotional skills, a recent review (Clarke et al. 2015) on programs for improving adolescents' social and emotional skills shows that the existing programs suffer from several problems and shortcomings. School-based programs do not easily integrate with adolescents' personal lives and community context. Meanwhile, out-of-school programs, including art-based programs, often lack rigorous, well-conducted evaluation of the program outcomes, and there is often insufficient theoretical grounding of programs on scientific knowledge about the underlying impact mechanisms. Music-based programs hold great potential regarding direct integration to the personal lives of individuals, and the use of solid theoretical constructs such as social-emotional competence as a basis for their design and evaluation can significantly improve program validity testing.

Conclusion

There is a close link between music, emotion, and social well-being. Music can be a tremendously rich resource not only for the development of social-emotional skills in daily life and through prevention/intervention programs but also for the observation and study of humans as social and emotional beings. Because music is rooted in the evolution of human nature, there are many aspects of socialization, communication, affect, and expressivity that have been sustained by and developed together with music (Cross and Morley 2009; McDermott and Hauser 2005; Miranda et al. 2015). As research has been pointing out, music accompanies us throughout all our life, and it can support both our weaknesses and strengths.

The word *forum* represents this aspect of music quite well—music creates a space, both private and social, where competences are worked and information is shared. Despite the close relations between music uses and well-being, it is hard to pin point causalities. However, as with every human activity, we might speculate that music engagement equally reflects and shapes social-emotional skills.

The pioneering research that we have been observing to emerge will possibly contribute to a better understanding in the future of the directions of the relations between music uses and social-emotional skills. We particularly argue for a need of future research to clarify the concept of *music-related* social-emotional competence. Music research needs constant dialogue with general psychological research on SEC, but it also needs to develop an understanding that is grounded in music, in order to identify which subcomponents of SEC are best supported by music in particular and what are the conditions under which music optimally serves as a promoter of these skills. The transference of this kind of knowledge to prevention and intervention programs will most definitely prove to be fruitful for the promotion of a stronger mental health among our youth and communities.

References

Altenmuller, E. K., Schurmann, V. K., & Parlitz, D. (2002). Hits to the left, flops to the right: Different emotions during listening to music are reflected in cortical lateralisation patterns. *Neuropsychologia, 40*, 2242–2256.

Arnett, J. J. (1995). Adolescents' uses of media for self-socialization. *Journal of Youth and Adolescence, 24*(5), 535–549.

Baltazar, M., & Saarikallio, S. (2016). Toward a better understanding and conceptualization of affect self-regulation through music: A critical, integrative literature review. *Psychology of Music, 44*(6), 1500–1521.

Bar-On, R. (1997). *The emotional quotient inventory (EQ-i): A test of emotional intelligence*. Toronto: Multi-Health Systems.

Bar-On, R. (2006). The bar-on model of emotional-social intelligence (ESI). *Psicothema, 18*, 13–25.

Bartlett, D. L. (1996). Physiological responses to music and sound stimuli. In D. A. Hodges (Ed.), *Handbook of music psychology* (pp. 343–385). San Antonio: University of Texas.

Bigand, E., Vieillard, S., Madurell, F., Marozeau, J., & Dacquet, A. (2006). Multidimensional scaling of emotional responses to music: The effect of musical expertise and of the duration of the excerpts. *Cognition and Emotion, 19*, 1113–1139.

Blood, A. J., & Zatorre, R. J. (2001). Intensely pleasurable responses to music correlate with activity in brain regions implicated in reward and emotion. *Proceedings of the National Academy of Sciences of the United States of America, 98*(20), 11818–11823.

Born, G. (2013). Introduction – Music, sound and space: Transformations of public and private experiences. In G. Born (Ed.), *Music, sound and space: Transformations of public and private experiences* (pp. 1–70). Cambridge: Cambridge University Press.

Carlson, E., Saarikallio, S., Toiviainen, P., Bogert, B., Kliuchko, M., & Brattico, E. (2015). Maladaptive and adaptive emotion regulation through music: A behavioral and neuroimaging study of males and females. *Frontiers in Human Neuroscience, 9*, 466.

Charles, S. T., & Carstensen, L. L. (2010). Social and emotional aging. *Annual Review of Psychology, 61*(1), 383–409.

Chin, T. C., & Rickard, N. S. (2014). Emotion regulation strategy mediates both positive and negative relationships between music uses and well-being. *Psychology of Music, 42*(5), 692–713.

Clarke, E. F. (2013). Music, space, and subjectivity. In G. Born (Ed.) *Music, sound and space: Transformations of public and private experiences* (pp. 90–110). Cambridge: Cambridge University Press.

Clarke, A. M., Morreale, S., Field, C-A., Hussein, Y., Barry, M. M. (2015). What works in enhancing social and emotional skills development during childhood and adolescence? A review of the evidence on the effectiveness of school-based and out-of-school programmes in the UK. A report produced by the World Health Organization Collaborating Centre for Health Promotion Research, National University of Ireland Galway.

Clayton, M. (2009). The social and personal functions of music in cross-cultural perspective. In. S. Hallam, I. Cross, & M. Thaut (Eds.), *The Oxford handbook of music psychology* (pp. 35–44). Oxford: Oxford University Press.

Clift, S., & Morrison, I. (2011). Group singing fosters mental health and wellbeing: Findings from the East Kent 'Singing for Health' network project. *Mental Health and Social Inclusion, 15*(2), 88–97.

Cross, I. (2008). Musicality and the human capacity for culture. *Musicae Scientiae, 12* (1), Special Issue, 147–167.

Cross, I. (2014). Music and communication in music psychology. *Psychology of Music, 42*(6), 809–819.

Cross, I., & Morley, I. (2009). The evolution of music: Theories, definitions and the nature of the evidence. In S. Malloch, & C. Trevarthen (Eds.), *Communicative musicality* (pp. 61–82). Oxford: Oxford University Press.

Davidson, J., & Emberly, A. (2012). Embodied musical communication across cultures: Singing and dancing for quality of life and wellbeing benefit. In R. MacDonald, G. Kreutz, L. Mitchell (Eds.), *Music, health, and wellbeing* (pp. 137–149). Oxford: Oxford University Press.

Denham, S. A., Bassett, H., Mincic, M., Kalb, S., Way, E., Wyatt, T., & Segal, Y. (2012). Social–emotional learning profiles of preschoolers' early school success: A person-centered approach. *Learning and Individual Differences, 22*(2), 178–189.

DeNora, T. (1999). Music as a technology of the self. *Poetics, 27*(1), 31–56.

Dingle, G. A., & Fay, C. (2016). Tuned In: The effectiveness for young adults of a group emotion regulation program using music listening. *Psychology of Music*. https://doi.org/10.1177/0305735616668586.

Dissanayake, E. (2008). If music is the food of love, what about survival and reproductive success? *Musicae Scientiae, 12*(Special issue), 169–195.

Eisenberg, N., Zhou, Q., Liew, J., Champion, C., Pidada, S. (2006). Emotion, emotion-related regulation, and social functioning. In X. Chen, D. French, B. Schneider (Eds.), *Peer relationships in cultural context* (pp. 180–200). Cambridge: Cambridge University Press.

Evers, S., & Suhr, B. (2000). Changes of the neurotransmitter serotonin but not of hormones during short time music perception. *European Archives of Psychiatry and Clinical Neuroscience, 250*, 144–147.

Field, T., Martinez, A., Nawrocki, T., Pickens, J., Fox, N. A., & Schanberg, S. (1998). Music shifts frontal EEG in depressed adolescents. *Adolescence, 33*, 109–116.

Gabrielsson, A., & Lindström, E. (2010). The role of structure in the musical expression of emotions. In P. N. Juslin & J. A. Sloboda (Eds.), *Handbook of music and emotion: Theory, research, and applications* (pp. 367–400). New York: Oxford University Press.

Garnefski, N., Kraaij, V., & van Etten, M. (2005). Specificity of relations between adolescents' cognitive emotion regulation strategies and internalizing and externalizing psychopathology. *Journal of Adolescence, 28*(5), 619–631.

Garrido, S., & Schubert, E. (2013). Adaptive and maladaptive attraction to negative emotions in music. *Musicae Scientiae, 17*(2), 147–166.

Gembris, H. (2008). Musical activities in the third age: An empirical study with amateur musicians. In *Proceedings of the 2nd European Conference on Developmental Psychology of Music*, London, UK, 10–12 September.

Gold, C., Crooke, A. H. D., Saarikallio, S., & McFerran, K. (2017). Group music therapy as a preventive intervention for young people at risk: Cluster-randomized trial. *Journal of Music Therapy*. https://doi.org/10.1093/jmt/thx002.

Gross, J. J., & John, O. P. (2003). Individual differences in two emotion regulation processes: Implications for affect, relationships, and well-being. *Journal of Personality and Social Psychology, 85*(2), 348–362.

Hausmann, M., Hodgetts, S., & Eerola, T. (2016). Music-induced changes in functional cerebral asymmetries. *Brain and Cognition, 104*, 58–71.

Hays, T., & Minichiello, V. (2005). The meaning of music in the lives of older people: A qualitative study. *Psychology of Music, 33*(4), 437–451.

Hughes, C., White, A., Sharpen, J., & Dunn, J. (2000). Antisocial, angry, and unsympathetic: "Hard-to-manage" preschoolers' peer problems and possible cognitive influences. *Journal of Child Psychology & Psychiatry, 41*, 169–179.

Jones, D. E., Greenberg, M., & Crowley, M. (2015). Early social-emotional functioning and public health: The relationship between kindergarten social competence and future wellness. *American Journal of Public Health, 105*(11), 2283–2290.

Juslin, P. N. (1997). Emotional communication in music performance: A functionalist perspective and some data. *Music Perception, 14*, 383–418.

Juslin, P. N., & Laukka, P. (2004). Expression, perception, and induction of musical emotions: A review and a questionnaire study of everyday listening. *Journal of New Music Research, 33*(3), 217–238.

Juslin, P. N., & Timmers, R. (2010). Expression and communication of emotion in music. In P. N. Juslin & J. Sloboda (Eds.), *Handbook of music and emotion: Theory, research, applications* (pp. 453–489). New York: Oxford University Press.

Juslin, P.N. & Sloboda, J. A. (Eds.) (2001). Music and emotion: Theory and research. New York: Oxford University Press.

Keane, S. P., & Calkins, S. D. (2004). Predicting kindergarten peer social status from toddler and pre-school problem behavior. *Journal of Abnormal Child Psychology, 32*, 409–423.

Kirschner, S., & Tomasello, M. (2010). Joint music making promotes prosocial behavior in 4-year-old children. *Evolution and Human Behavior, 31*, 354–364.

Koelsch, S. (2014). Brain correlates of music-evoked emotions. *Nature Reviews Neuroscience, 15*(3), 170–180.

Koelsch, S., & Stegemann, T. (2012). The brain and positive biological effects in healthy and clinical populations. In R. MacDonald, G. Kreutz, L. Mitchell (Eds.), *Music, health, and wellbeing* (pp. 436–456). Oxford: Oxford University Press.

Koelsch, S., Boehlig, A., Hohenadel, M., Nitsche, I., Bauer, K., & Sack, U. (2016). The impact of acute stress on hormones and cytokines, and how their recovery is affected by music-evoked positive mood. *Science Reports, 6*, 23008.

Kreutz, G., Quiroga Murcia, C., Bongard, S. (2012). Psychoneuroendocrine research on music and health: An overview. In R. MacDonald, G. Kreutz, L. Mitchell (Eds.), *Music, health, and wellbeing* (pp. 457–476). Oxford: Oxford University Press.

Krueger, J. (2014). Affordances and the musically extended mind. *Frontiers in Psychology, 4*(1003), 1–13.

Laiho, S. (2004). The psychological functions of music in adolescence. *Nordic Journal of Music Therapy, 13*(1), 49–65.

Laukka, P. (2006). Uses of music and psychological well-being among the elderly. *Journal of Happiness Studies, 8*(2), 215–241.

Lee, C. H., Lee, C. Y., Hsu, M. Y., Lai, C. L., Sung, Y. H., Sung, Y. H., & Lin, C. Y. (2017). Effects of music intervention on state anxiety and physiological indices in patients undergoing mechanical ventilation in the intensive care unit: A randomized controlled trial. *Biological Research for Nursing, 19*, 137–144. https://doi.org/10.1177/1099800416669601.

Lincoln, S. (2005). Feeling the noise: Teenagers, bedrooms, and music. *Leisure Studies, 24*(4), 399–414.

MacDonald, R., Hargreaves, D. J., Miell, D. (2017). *Handbook of musical identities*. Oxford: Oxford University Press.

Malloch, S. N. (1999). Mothers and infants and communicative musicality. *Musicae Scientiae, 3*(Special issue), 29–57.

Marik, M., & Stegemann, T. (2016). Introducing a new model of emotion dysregulation with implications for everyday use of music and music therapy. *Musicae Scientiae, 20*(1), 53–67.

Mayer, J. D., & Salovey, P. (1997). What is emotional intelligence? In P. Salovey & D. Sluyter (Eds.), *Emotional development and emotional intelligence: Implications for educators* (pp. 3–31). New York: Basic Books.

McDermott, J., & Hauser, M. (2005). The origins of music: Innateness, uniqueness, and evolution. *Music Perception, 23*(1), 29–60.

McFerran, K. S. (2016). Contextualising the relationship between music, emotions and the well-being of young people: A critical interpretive synthesis. *Musicae Scientiae, 20*(1), 103–121.

Menon, V., & Levitin, D. (2005). The rewards of music listening: Response and physiological connectivity of mesolimbic system. *NeuroImage, 228*, 175–184.

Miranda, D. (2013). The role of music in adolescent development: Much more than the same old song. *International Journal of Adolescence and Youth, 18*(1), 5–22.

Miranda, D., Blais-Rochette, C., Vaugon, K., Osman, M., & Arias-Valenzuela, M. (2015). Towards a cultural-developmental psychology of music in adolescence. *Psychology of Music, 43*(2), 197–218.

Mithen, S. (2005). *The singing Neanderthals: The origin of music, language, mind, and body*. London: Weidenfeld & Nicholson.

Mörelius, E., He, H. G., & Shorey, S. (2016). Salivary cortisol reactivity in preterm infants in neonatal intensive care: An integrative review. *International Journal of Environmental Research and Public Health, 13*(3), 337. https://doi.org/10.3390/ijerph13030337.

Murcia, C. Q., & Kreutz, G. (2012). Dance and health: Exploring interactions and implications. In R. MacDonald, G. Kreutz, L. Mitchell (Eds.), *Music, health, and wellbeing* (pp. 126–135). Oxford: Oxford University Press.

North, A. C., Hargreaves, D. J., & O'Neill, S. A. (2000). The importance of music to adolescents. *British Journal of Educational Psychology, 70*, 255–272.

Oldfield, A. (1993). *Interactive music therapy in child and family psychiatry: Clinical practice, research, and teaching*. London: Jessica Kingsley Publishers.

Pascual, A., Conejero, S., & Etxebarria, I. (2016). Coping strategies and emotion regulation in adolescents: Adequacy and gender differences. *Ansiedad y Estrés, 22*, 1–4.

Payton, J. W., Wardlaw, D. M., Graczyk, P. A., Bloodworth, M. R., Tompsett, C. J., & Weissberg, R. P. (2000). Social and emotional learning: A framework for promoting mental health and reducing risk behaviors in children and youth. *Journal of School Health, 70*, 179–185.

Petrides, K. V., Niven, L., & Mouskounti, T. (2006). The trait emotional intelligence of ballet dancers and musicians. *Psicothema, 18*(supl), 101–107.

Randall, W. M., Rickard, N. S., & Vella-Brodrick, D. A. (2014). Emotional outcomes of regulation strategies used during personal music listening: A mobile experience sampling study. *Musicae Scientiae, 18*(3), 275–291.

Rose-Krasnor, L. (1997). The nature of social competence: A theoretical review. *Social Development, 6*(1), 111–135.

Ruud, E. (1997). Music and the quality of life. *Nordic Journal of Music Therapy, 6*(2), 86–97.

Saarikallio, S. (2017). Musical identity in fostering emotional health. In R. MacDonald, D. J. Hargreaves, and D. Miell (Eds.), *Handbook of musical identities* (pp. 602–623), Oxford: Oxford University Press.

Saarikallio, S., & Erkkilä, J. (2007). The role of music in adolescents' mood regulation. *Psychology of Music, 35*(1), 88–109.

Saarikallio, S., Nieminen, S., & Brattico, E. (2013). Affective reactions to musical stimuli reflect emotional use of music in everyday life. *Musicae Scientiae, 17*(1), 27–39.

Saarikallio, S., Vuoskoski, J., & Luck, G. (2014). Adolescents' expression and perception of emotion in music reflects their broader abilities of emotional communication. *Psychology of Well-Being, 4*, 21.

Saarikallio, S., McFerran, K. S., & Gold, C. (2015). Development and validation of the healthy-unhealthy music scale (HUMS). *Child and Adolescent Mental Health, 20*, 210–217.
Saarni, C. (1990). Emotional competence: How emotions and relationships become integrated. In R. A. Thompson (Ed.), *Nebraska symposium on motivation* (Vol. 36, pp. 115–182). Lincoln: University of Nebraska Press.
Saarni, C. (1999). *The development of emotional competence*. New York: Guilford.
Schmidt, L. A., & Trainor, L. J. (2001). Frontal brain electrical activity (EEG) distinguishes valence and intensity of musical emotions. *Cognition & Emotion, 15*(4), 487–500.
Seiffge-Krenke, I. (2000). Causal links between stressful events, coping style, and adolescent symptomatology. *Journal of Adolescence, 23*, 675–691.
Selfhout, M. H. W., Branje, S. J. T., ter Bogt, T. F. M., & Meeus, W. H. J. (2009). The role of music preferences in early adolescents' friendship formation and stability. *Journal of Adolescence, 32*(1), 95–107.
Sifneos, P. E. (1996). Alexithymia: Past and present. *The American Journal of Psychiatry., 153*(7 Suppl), 137–142.
Silk, J. S., Steinberg, L., & Morris, A. S. (2003). Adolescents' emotion regulation in daily life: Links to depressive symptoms and problem behavior. *Child Development, 74*(6), 1869–1880.
Sloboda, J. A., & Juslin, P. N. (2010). At the interface between the inner and outer world: Psychological perspectives. In Patrik N. Juslin, & John A. Sloboda (Eds.), *Handbook of music and emotion: Theory, research, and applications* (73–98). New York: Oxford University Press.
Sloboda, J. A., & O'Neill, S. A. (2001). Emotions in everyday listening to music. In Patrik N. Juslin, & John A. Sloboda (Eds.), *Music and emotion: Theory and research* (pp. 415–429). New York: Oxford University Press.
Steinberg, L. (2005). Cognitive and affective development in adolescence. *Trends in Cognitive Sciences, 9*(2), 69–74.
Tafuri, J., & Villa, D. (2002). Musical elements in the vocalisations of infants aged 2–8 months. *British Journal of Music Education, 19*(1), 73–88.
Thayer, R. E., Newman, J. R., & McClain, T. M. (1994). Self-regulation of mood: Strategies for changing a bad mood, raising energy, and reducing tension. *Journal of Personality and Social Psychology, 67*(5), 910–925.
Theorell, T. P., Lennartsson, A. K., Mosing, M. A., & Ullén, F. (2014). Musical activity and emotional competence – a twin study. *Frontiers in Psychology, 5*, 774.
Thoma, M. V., Ryf, S., Mohiyeddini, C., Ehlert, U., & Nater, U. M. (2012). Emotion regulation through listening to music in everyday situations. *Cognition and Emotion, 26*(3), 550–560.
Thoma, M. V., La Marca, R., Brönnimann, R., Finkel, L., Ehlert, U., & Nater, U. M. (2013). The effect of music on the human stress response. *PLoS One, 8*(8), e70156. https://doi.org/10.1371/journal.pone.0070156.
Thompson, W. F., Schellenberg, E. G., & Husain, G. (2004). Decoding speech prosody: Do music lessons help? *Emotion, 4*(1), 46–64.
Thomson, C., Reece, J., & Di Benedetto, M. (2014). The relationship between music-related mood regulation and psychopathology in young people. *Musicae Scientiae, 18*(2), 150–165.
Trainor, L. J., & Cirelli, L. (2015). Rhythm and interpersonal synchrony in early social development. *Annals of the New York Academy of Sciences, 1337*(1), 45–52.
Trehub, S. E., Trainor, L. J., & Unyk, A. M. (1993). Music and speech processing in the first year of life. *Advances in Child Development and Behavior, 24*, 1–35.
Trevarthen, C. (2002). Origins in musical identities: Evidence from infancy for musical social awareness. In R. MacDonald, D. J. Hargreaves, D. Miell (Eds.), *Handbook of musical identities* (pp. 28–33). Oxford: Oxford University Press.
Van den Tol, A. J. M., & Edwards, J. (2014). Listening to sad music in adverse situations: How music selection strategies relate to self-regulatory goals, listening effects, and mood enhancement. *Psychology of Music, 43*(4), 473–494.
Van Goethem, A. (2010). Affect regulation in everyday life: Strategies, tactics, and the role of music. Unpublished Dissertation. University of Keele.

Vieillard, S., Peretz, I., Gosselin, N., Khalfa, S., Gagnon, L., & Bouchard, B. (2008). Happy, sad, scary, and peaceful musical excerpts for research on emotions. *Cognition and Emotion, 22*, 720–752.

Wöllner, C. (2012). Is empathy related to the perception of emotional expression in music? A multimodal time-series analysis. *Psychology of Aesthetics, Creativity, and the Arts, 6*(3), 214–223.

World Health Organization (WHO) (2004). Promoting mental health: Concepts, emerging evidence, practice – summary report. Online. Available at http://www.who.int/mental_health/evidence/en/promoting_mhh.pdf.

Zentner, M., Grandjean, D., & Scherer, K.R. (2008). Emotions evoked by the sound of music: Characterization, classification, and measurement. *Emotion 8*(4):494–521.

Chapter 8
Partnerships for Health Musicking: A Case for Connecting Music Therapy and Public Health Practices

Brynjulf Stige

Introduction

There are several usages of the notion of public health, so I will position myself by putting forward three statements. First, the notion of public health used in this chapter includes but goes beyond population-oriented prevention of disease to include healthy public policy that can create supportive environments and promote positive aspects of health. Second, such healthy environments include the health-care system as well as other sectors in society such as transport, work, education and culture, so that it is highly relevant to connect practices in clinics and in broader communities. Third, public health strategies are based in the human rights, so that social justice, participation and empowerment are integral rather than additional goals (see Chap. 9 by Stensæth). These statements will be developed throughout the chapter and find support in the World Health Organization's (1986, 1998) formulations of new public health strategies for the twenty-first century, where, for instance, health promotion is understood as the process of enabling people to increase control over, and to improve, their health.[1]

Textbooks on music therapy often do not discuss music and public health very thoroughly. In the most recent edition of Bruscia's influential text *Defining Music Therapy*, for instance, the notion of public health is mentioned only once and just briefly, with reference to the use of music to facilitate public health education. This, then Bruscia (2014, Chap. 21) considers an auxiliary practice to music therapy. The judgement is partly understandable, because there is a tradition for thinking about therapy and public health as unrelated and sometimes even competing practices

[1] In this chapter, I provide practice examples from the Norwegian context where I work, so it is also worth noting that these statements find support in the Norwegian law for public health work (Lov om folkehelsearbeid 2011).

B. Stige (✉)
The Grieg Academy Music Therapy Research Centre, University of Bergen, Bergen, Norway
e-mail: Brynjulf.Stige@uib.no

L. O. Bonde, T. Theorell (eds.), *Music and Public Health*,
https://doi.org/10.1007/978-3-319-76240-1_8

(Turner 2004). Notwithstanding the many other qualities in Bruscia's text, I still contend that the lack of discussion of relationships between music therapy and public health reflects a limiting conception of music therapy.

Having said this, I am myself guilty of having co-authored a textbook on music therapy that only briefly touches upon the issue of public health (Bunt and Stige 2014). In contrast, I have co-authored a textbook on community music therapy where the notion of public health is brought up a number of times and seen in relation to issues such as poverty and social justice, positive health and wellbeing, participation and social capital and cross-professional and cross-sectorial collaboration (Stige and Aarø 2012).[2] This could be taken to suggest that public health issues are of limited relevance to the field of music therapy as a whole, although it is relevant to socially oriented subfields such as community music therapy. I would rather argue that community music therapy is one of several recent movements that support the idea that will be discussed in this chapter, namely, that it is about time that we re-examine relationships between music therapy and public health more carefully.

I will try to contribute in this direction through elaboration of the notions of *health musicking* and *partnership*, which will be used to contextualize the request for user involvement and collaboration that in some ways and to some degree counters the current tendency to consider health care as industry and business. Developments within health-care services often share many characteristics from one country to the next, but they are also situated in particular political, socioeconomic and cultural contexts. I will therefore start with some reflections on how WHO initiatives on public health and health promotion have affected music therapy in the Norwegian context where I work myself.

Alma Ata and Ottawa Revisited: Health Promotion Within and Without the Clinic

The Alma Ata conference set up by WHO in 1978 produced a declaration that highlighted that health is a fundamental human right:

> The Conference strongly reaffirms that health … is a fundamental human right and that the attainment of the highest possible level of health is a most important world-wide social goal whose realization requires the action of many other social and economic sectors in addition to the health sector. (World Health Organization 1978)

Some years later, the Ottawa charter for health promotion described the following prerequisites for health:

> The fundamental conditions and resources for health are peace, shelter, education, food, income, a stable eco-system, sustainable resources, social justice and equity. Improvement in health requires a secure foundation in these basic prerequisites. (World Health Organization 1986)

---

[2] Community music therapy is a socio-musical movement within music therapy, with qualities that could be described by the acronym PREPARE; practices are *p*articipatory, *r*esource-oriented, *e*cological, *p*erformative, *a*ctivist, *r*eflective and *e*thics-driven (Stige and Aarø 2012, pp. 3–28).

In other words, our health depends on healthy public policy. Consequently, action for health promotion involves all sectors of society and goes far beyond the mandates of health professionals and the health-care system.

In the 1980s, when these documents were subject to public debate, community music therapy practices were emerging in Norway, with a focus upon every person's right to cultural participation (Kleive and Stige 1988). The documents from Alma Ata and Ottawa provided a new context for reflecting upon these developments of rights-based practice, by clarifying relationships between health practices and human rights. Consequently, they challenged the thinking of this author and other Norwegian music therapists struggling to understand relationships between didactic, cultural and clinical practices of music therapy.

Theoretical implications include the development of ecological and participatory perspectives on music therapy (Stige 1996, 2002), but only recently has it in my country been possible to explore more thoroughly what the implications for service development could be. Such opportunities were created a few years ago, when the Norwegian Health Directorate included a strong recommendation of music therapy in the guidelines for treatment of persons with psychotic disorders (Helsedirektoratet 2013). These guidelines represent a legitimate platform for implementing music therapy in all clinics and municipalities nationally. With this opportunity for the profession, there is a serious obligation as well; the Norwegian system of publicly funded health care is based in the social justice principle that all citizens should have equal rights to health-care services. Such equal access in no way exists when it comes to music therapy today, which is a moral dilemma for Norwegian music therapists as well as for health leaders and politicians.

In the area around the city of Bergen, where I currently work, the response in 2013 was to start developing a knowledge cluster bringing university researchers, practitioners, service users and leaders together, which I will describe later in the article. Here, I will exemplify developments by describing the strategies for implementing music therapy within mental health services in Bergen Health Trust (Helse Bergen), a subdivision of the Western Norway Regional Health Authority (Helse Vest). This trust is responsible for six community mental health clinics, which together cover an area with a population of several hundred thousand people, almost one out of ten of the nation's total population of about five million people.

Until very recently, only one of these six clinics had employed a music therapist, and there were no overall plan for implementing music therapy systematically. In 2015 and 2016, things started to change, however, with a new requirement from the government. All health trusts were instructed to develop medication-free services to patients preferring such treatment. The original initiative for medication-free treatment came from user organizations nationally. Not only the option of *not* taking such medication should be available but also high-quality psychosocial alternatives, the user organizations have argued for years.[3]

---

[3]The public health department of the Norwegian Health Directorate has created a website which describes users' rights to medication-free treatment within mental health services. The website includes several links to websites with information about music therapy; see https://helsedirektoratet.no/folkehelse/psykisk-helse-og-rus/psykisk-helsevern/legemiddelfri-behandling-i-psykisk-helsevern.

In Bergen, local partnerships were established to implement the new national policy. The chosen model for medication-free treatment in our region involves shared decision-making in relation to a range of available services, where music therapy is now added to the list. This creates, for the first time, possibilities for systematic implementation of music therapy in the specialized mental health services in the region. Then, in January 2017, Bergen Health Trust developed a new health promotion strategy that stated that all clinics should have one or more music therapists.

To my knowledge, this is the first health trust in Norway with systematic implementation of music therapy services. When asked, a representative of Bergen Health Trust explains the decision by referring to the disturbing public health fact that people with serious mental health challenges have an expected lifespan that is more than 20 years shorter than the average for the total population. There are several reasons for this, the representative explains, including drugs and medication and lifestyle issues such as diet and physical activity. He continues by explaining that the health trust has an obligation to employ measures across a wide range in an attempt of alleviating the situation and that music therapy was included in the strategy both because it was considered an evidence-based treatment and a promising part of a health promotion strategy that could empower patients and mobilize resources (Geir Lien in POLYFON-nytt 2017).

Complementing this strategy of having music therapists in each clinic, the health trust also supports a project in a local community arts centre. In the project, called MOT82, music therapists work to support and empower participants who previously used mental health services in one of the clinics of the health trust. The project is led by music therapist Lars Tuastad, and it gives support to the participants' process of gaining access to cultural activities and of making music a self-monitored health promoting daily life activity. In this process, collaboration with local musicians, organizations and authorities of culture is central.[4]

In a very concrete way, this project exemplifies one attempt of reorienting the health sector in a health promotion direction, in line with suggestions made in the Ottawa Charter:

> The role of the health sector must move increasingly in a health promotion direction, beyond its responsibility for providing clinical and curative services. Health services need to embrace an expanded mandate which is sensitive and respects cultural needs. This mandate should support the needs of individuals and communities for a healthier life, and open channels between the health sector and broader social, political, economic and physical environmental components. (World Health Organization 1986, p. 3)

In order to contextualize the developments described above – where public health objectives, music therapy and health promotion activities are connected – I will outline the notion of health musicking.

[4] In Norwegian, there is a description of the project MOT82 in the following website: http://gamut.no/2017/03/07/onsker-a-hjelpe-flere-mennesker-med-psykiske-lidelser-til-a-delta-i-lokale-aktiviteter/.

Health Musicking

I have previously argued that the discipline of music therapy could be defined as the study and learning of relationships between music and health (Stige 2002, p. 198). This definition of the discipline supports a broad conceptualization of music therapy research and scholarship, to include various health-related practices of music. Music therapy should then be able to contribute to an interdisciplinary discourse, such as the public health discourse, which is also relevant for the development of new perspectives on collaborative music therapy practices across context, as exemplified above.

Inspired by Wittgenstein's (1953/1967) perspective on meaning-making as participation in situated activity, Small's (1998) concept of music as situated activity and DeNora's (2000) discussion of music and action, I have developed the notion of *health musicking* to communicate the idea that relationships between music and health could be understood as situated processes of participation. Such processes evolve inside and outside conventional music therapy practices (Stige 2002, 2006, 2012). Originally, I defined health musicking as the appraisal and appropriation of the health affordances of the arena, agenda, agents, activities and artefacts of a music practice (Stige 2002, p. 211):

Arena: Musicking is a situated activity, linked to a site and situation, and health could be described similarly. There are several sites of interaction to take into consideration, then, such as body, person, dyad, group, organization, and locality. The resources of an arena can be manifold and provide a foundation for the activities that agents take part in. Consequently, an arena is also a site of struggle, between various interests and values.

Agenda: The evolving issues and goals that participants relate to, consciously and unconsciously, constitute the agenda of the activity. Agendas are negotiated more or less openly, and this process can be valuable and/or problematic, by affording communication, collaboration and conflict, for instance.

Agents: As used here, the term *agents* particularly refers to the human actors involved in the activity. Agents experience agency to a higher or lower degree, that is, they can be empowered to experience that they have the capacity to influence the flow and direction of the activity. Sometimes agents form alliances in order to enhance their agency, and such alliances can be more or less inclusive, more or less democratic and so on.

Activities: Broadly speaking, the term activity refers to interactions between organism and environment. More specifically, this includes engagement with music: *listening*, *playing*, *creating*, *performing*, *interpreting* and *reflecting*. As used here, *interpreting* involves translation of the sounds of music through use of another modality (such as movement, art or poetry), while *reflecting* puts the other activities in perspective, usually through verbal or written processing but also through use of other available modalities.

Artefacts: Health musicking may involve use of several different artefacts, for instance, musical instruments, musical notation and lyrics and various forms of recording equipment. The theory traditions that inform the notion suggest that a person's sense of self is constituted through internalization and creative use of cultural artefacts in various social contexts. An example would be: Musical instruments can invite playing and thus participation in a social context, which invites responses from and collaboration with others, which provides experiences that influence one's sense of identity and so on.

For all the components that I have described above, there are reciprocal processes of shaping involved. An arena allows certain agendas to be negotiated, provides artefacts, affords certain activities and so on. For instance, the mental health clinic described above provides some of the same artefacts and activities as those provided by the local arts centre and the MOT82 project and some artefacts and activities that differ. It would be possible to participate in a music therapy rock band in both arenas, for instance, while the mental health clinic has better facilities for individual therapy processes and the arts centre better facilities for music café events. In the same way, agendas may overlap or vary, and they evolve over time. In the MOT82 project at the arts centre, it is more common than in the clinic to examine the possibilities of developing music as a self-monitored everyday activity, for instance.

The terms offered here for description of music as health-related activity are developed within a perspective where *participation* is considered central to human development and wellbeing. This is compatible with a bioecological (Bronfenbrenner 1979, 2005) and cultural psychology (Cole 1996; Heine 2015) perspectives on "how humans become human." In no way such sociocultural perspectives should be taken to ignore biological aspects of human development, but less in the direction of sociobiology (Wilson 1975/2000) and more in the direction of neurosociology (Franks and Turner 2013). The sociocultural matrixes of the mutually attuned and coordinated bodies that our brains are part of should be taken into consideration.

Partnerships for Health

In describing health musicking above, I clarified that participating agents sometimes form alliances in order to enhance their agency. In the public health literature, alliances that are characterized by trust, collective decision-making and dedication to shared goals are often labelled *partnerships* (World Health Organization 1998).

Causes that bring agents together are often complex and multifaceted, so that each partner realizes that collaboration across established lines of differentiation is helpful. Partnerships for health might involve agents in a range of sectors, working on various levels. An almost infinite diversity of goals and actions is imaginable (Amdam 2010).

Four types of partnerships that might be particularly relevant for music therapy are self-help partnerships, project partnerships, institutional partnerships and governance partnerships. *Self-help partnerships* for lifestyle changes and social change in a given context are collaborative relationships characterized by a democratic process of collective decision-making. Local and informal partnerships geared towards shared goals of say service development are often called *project partnerships*. Formalized collaboration between institutions is often called *institutional partnerships*, which are more judicially binding than project partnerships. In *governance partnerships*, public interests collaborate innovatively with private interests and/or the third sector (civil society) in order to realize shared policy goals (Stige and Aarø 2012, pp. 274–278).

Different types of partnerships are not mutually exclusive. A governance partnership, for instance, could stimulate self-help partnerships as well as project partnerships and institutional partnerships. Given the fact that public health work requires healthy public policy, governance partnerships are worth particular attention, and I will elaborate a little bit here upon this type of partnership.

In a study on governance for health in the twenty-first century, the World Health Organization (2012) examines various governance innovations, including joint action of various sectors in society and of public, private and civil society actors as well. Framed in this way, governance for health can be collaborative, similar to what Amdam (2010) calls governance partnerships where public authorities collaborate with a range of agents in order to realize shared goals. Governance partnerships or collaborative governance is often described as democratic, in that it is based on consensus-oriented decision-making, but the analysis performed by Ansell and Gash (2007) indicates that outcomes depend on critical variables such as prior history of conflict or cooperation, incentives for participation and power and resources imbalances. These authors also identify factors that are central in such collaborative processes, including face-to-face dialogue, trust building and the development of commitment and shared understanding.

Governance partnerships are, for instance, relevant as collaborative efforts for implementation of new knowledge, and for reorienting health services, in line with suggestions made in the Ottawa Charter (see above). Despite an increasing amount of primary and review studies in the field, it still remains uncertain how and under what conditions change strategies and interventions most effectively can be translated and exchanged to health professionals and integrated in their organizations (Pentland et al. 2011; Flodgren et al. 2011, 2012). Recent case and framework studies have been able to identify, however, that major factors in successful knowledge application are organizational, with leaders playing a vital role in creating and supporting their organizations' process of knowledge implementation (Berta et al. 2010; Mekki 2015; Ward et al. 2012).

Therefore, to see how governance is managed and carried out in health-care services and public health work is needed. In such processes, partnerships might represent alternatives to vertical hierarchies and horizontal demarcations, but they are not without their own limitations and problems. To overcome differences in power or histories of mistrust might be quite challenging, for instance. Consequently, partnerships might run the risk of growing unstable and vulnerable, unless one is able to address and negotiate such issues successfully. The idea of partnerships could also be misused, for instance, as a weasel word when power and privilege have not been substantially challenged. Partnerships do not automatically erase conflict and competition, even though they might create a new and different frame for communication and negotiation (Amdam 2010).

In the following, I will present one example of a governance partnership, established in Norway in 2015, in response to the fact that national guidelines did recommend music therapy while implementation was fragmented and unsystematic, in spite of the fact that Norwegian citizens by law have a right to equal access to services.

POLYFON Knowledge Cluster for Music Therapy

Since 2013, music therapy has been recommended in several national policy documents in Norway (e.g. Helsedirektoratet 2013), partly with reference to the evidence on the therapeutic effects of music therapy, partly with reference to music therapy's capacity to promote user involvement and participation. Such conceptual distinctions are not always clear, however, and perhaps the documents could be said to disguise conflicting narratives (Jacobsen 2015), where support for innovative and humane care is drawn from narratives of evidence and health economy, with critical examination of current research wanting. This situation both creates a need to implement current knowledge and to examine the status of this knowledge. With Miller et al. (2012), we could perhaps say that there is a need for a critical ecology than can nurture participation, professionalization and critical reflection.

The planning of POLYFON knowledge cluster for music therapy started in 2013 when the first national treatment guidelines recommending music therapy were published. The cluster was established in January 2015, and in 2017 the partners have committed to a second period of collaboration that lasts through 2020. POLYFON is coordinated by the Grieg Academy at the University of Bergen, and it is established with the goal of developing more and better music therapy services, education, research and dissemination.

There are more than ten partners, including three research centres, hospitals, health trusts, municipalities and a county. Most of the partners are public institutions, while some are non-profit health-care organizations, and one is a for-profit health-care company. The process of collaboration has revealed that these partners need each other; the health-care services need access to research information and to implementation research in order to initiate systematic implementation of a new practice such as music therapy, and the university and the research centres need collaborators in the field in order to develop solid research, education and dissemination activities of relevance to society.

When we chose POLYFON as the name of the partnership, we thought of it as a metaphor for a collaborative practice where different views could be voiced and related to each other.[5] In Norwegian, the name also works as an acronym: The Y in the centre refers to development of a new professional role ("yrkesrolle") that could

[5] "Polyfon" means *polyphonic* in Norwegian.

nurture an ecology of change at the levels of person, organisation and local community (POL) and be nurtured by partnerships involving civil society, the public sector and the private sector (FON).[6] The interplay between civil society, the public sector and the private sector is central in POLYFON, because the Nordic political model is based in the premise that these sectors should support, check and balance each other (Barth et al. 2003).

Areas of practice in POLYFON include child and adolescent development and welfare, mental health services, treatment of substance use problems, care for older adults and palliative care. Each area has established a cross-professional working group collaborating with user representatives. There is also an interdisciplinary Scientific Advisory Committee, with prominent international researchers from fields such as music sociology, music therapy, substance use problems and community mental health.[7]

The Proof of the Pudding

As described earlier in the chapter, the participation in the POLYFON knowledge cluster has enabled Bergen Health Trust to systematically implement music therapy in its mental health clinics. I consider it probable that in the next few years, health trusts around the country will make similar decisions. This is important and necessary in a social justice perspective, given that health is a human right (universally) and music therapy a recommended mental health service (in Norway). The degree to which this development is linked to a public health agenda is critical, however, because the guidelines that recommend music therapy are not only describing it as effective treatment but are also highlighting human rights issues such as involvement, shared decision-making and community participation (Helsedirektoratet 2013). While equal access to services is a necessary condition, the "proof of the pudding" is always in the eating. So, we need to examine the *use* of music therapy services, in the clinics and in the community, and our appraisal needs to include human rights issues such as participation and public health challenges such as social isolation and marginalization.

These developments are too recent for any systematic evaluation, but we can learn from the project MOT82. As described above, Bergen Health Trust has supplemented its strategy of having music therapists in each clinic with support to this project, which is led by music therapist Lars Tuastad and located in a local community arts centre.[8] The project provides participants with access to community

[6] Norwegian terms for civil society, the public sector and the private sector are "frivillig sektor", "offentleg sektor" and "næringsliv", respectively.

[7] There is a website (in Norwegian) established for the knowledge cluster, with information, annual reports and so on; see http://gamut.no/polyfon/.

[8] The project is co-funded by the Western Norway Regional Health Authority, Bergen Health Trust, the city of Bergen and the foundation ExtraStiftelsen.

music therapy services if needed, and a central function of the project is to provide support to participants' process of gaining access to socio-musical resources in the local community and of making music a self-monitored health-promoting activity in their everyday life.

After about 1 year, user experiences of the project were evaluated (Bjotveit et al. 2016). All 20 participants that used the services in the summer of 2016 agreed to respond to a survey about their experiences. The survey provided multiple-choice alternatives to most questions but also space for the participants' own descriptions of their experiences. Participants were invited to reflect upon how often they wanted to come to the arts centre for music and music therapy, which activities they wanted to engage with and so on. They were also invited to reflect upon how participation in music affected their everyday life experiences.

Preferences varied, as exemplified by the range of activities that participants were interested in: ten participants wanted to learn an instrument, nine wanted to play in a band, eight wanted to take part in a music discussion group, eight wanted to join the weekly music café, seven wanted to record in a studio, seven wanted to write songs, six wanted to take part in music listening activities, six wanted to join concerts, six wanted to learn to sing, six wanted to learn more about the theory of music, five wanted to produce music in a studio, four wanted to dance, three wanted to sing in a choir, two wanted to take part in a drum circle, two wanted to work with lyrics analysis and two wanted to play in a marching band (Bjotveit et al. 2016, p. 12).

While preferences will vary from place to place and group to group, the diversity demonstrated above should hardly surprise us. Participant preferences are coloured by cultural conditions but obviously also by individual characteristics and life histories. More striking, perhaps, is the fact that there are some issues where participant agreement is quite substantial. All 20 participants chose the "satisfied" or "very satisfied" alternatives when evaluating their experience of the activities they had taken part in, and 14 out of 20 found that the music activities had become a very important part of their everyday life. Also, a majority of the participants found that the activities allowed for user involvement in a satisfactory way (Bjotveit et al. 2016, pp. 13–21).

Seventeen out of 20 participants found that participation in the music activities made it easier to interact socially with other people, as illustrated by the following quotes:

> "[It's the] Only social [thing] I do during the week."
> "MOT82 has got me out of two years of isolation and allowed me to participate in other therapies."
> "Here you meet people with shared interests."
> "Music therapy balances negative emotions so they do not stand in the way of social interaction with other people."[9] (four different participants in MOT82, in Bjotveit et al. 2016, pp. 12–17)

When asked about whether or not the participants had any advice to give about future developments of music therapy and music activities in the community, one participant responded:

[9] These quotes are presented here in my translation of the Norwegian originals.

> Good therapy/follow-up costs a lot of money. In the long run, bad follow-up is a lot more expensive. You should dare to be adventurous, to think big, to think holistically, and to think in long-term perspectives! (If music therapy enables me to work; what does it cost to offer music therapy services versus the payment of disability benefits/hospital stays?) You do the math! (participant in MOT82, in Bjotveit et al. 2016, p. 26)

The math that the health authorities usually ask for is performed in the Cochrane Reviews and meta-analyses that inform the treatment guidelines where music therapy is recommended (e.g. Mössler et al. 2011; Gold et al. 2009). Increasingly also, we could expect the request for studies that calculate the health economics involved, as indicated in the participant quote above. In addition to the math, we need to bring forward the narratives of the people who use and experience the services, however. There are good reasons to suggest that the processes and effects of music therapy depend upon participants' use of music in context (Stige et al. 2010; Ansdell 2014), and studies of participant contributions will be key (Rolvsjord 2015).

Concluding Remarks

In a classic article within community psychology, Julian Rappaport stated that "Having rights but no resources and no services available is a cruel joke" (Rappaport 1981, p. 13). In this chapter I have described the development of music therapy services in mental health clinics as a resource for the realization of health as a human right, with practical implications for community and public health issues such as social isolation and marginalization.

Music therapy's relevance to public health depends upon several factors. Implementation of music therapy in the health-care sector is a necessary yet insufficient condition. Music therapists' ability to think and work across sectors and their capacity to embrace objectives such as positive health and community participation will be key. I have tried to develop a case for a theoretical understanding of music therapy, which highlights connections between clinical work and community work, with participation and collaboration as one of the common factors. The term *health musicking* was presented as one theoretical notion that could illuminate how situated participation is central to how music helps, within clinics and communities. The relevance of Bolger and Skewes McFerran's (2013) discussion of how music therapists need to care for the sustainability of practices towards self-supporting music projects follows.

I have used the term *partnership* to highlight the collaborative nature of processes that can increase music's availability as a health resource in society, and I have described the notion on several levels of analysis, from participant empowerment in self-help partnerships to the social change that governance partnerships might have the capacity to stimulate.

Perhaps in the future, it will be relevant to talk about *health musicianship* in ways similar to how health literacy today is used as a term to describe "the cognitive and social skills which determine the motivation and ability of individuals to gain access to, understand and use information in ways which promote and maintain good health" (World Health Organization 1998). The participants in MOT82 remind us, however, that music as an individualized health resource will only be part of the picture. Music is a social process; it invites community participation and allows for opportunities to challenge social isolation and marginalization. Health musicianship evolves through participation in the health musicking of communities of practice.

References

Amdam, R. (2010). *Planning in health promotion work. An empowerment model*. New York: Routledge.

Ansdell, G. (2014). *How music helps – in music therapy and everyday life*. Farnham: Ashgate.

Ansell, C., & Gash, A. (2007). Collaborative governance in theory and practice. *Journal of Public Administration Research and Theory, 18*, 543–571. https://doi.org/10.1093/jopart/mum032.

Barth, E., Moene, K. O. and Wallerstein, M. (2003). *Likhet under press: Utfordringer for den skandinaviske fordelingsmodellen* [Equality under Pressure. Challenges for the Scandinavian Model of Distribution]. Oslo: Gyldendal Norsk Forlag.

Berta, W., Teare, G. F., Gilbart, E., Ginsburg, L. S., Lemieux-Charles, L., Davis, D., et al. (2010). Spanning the know-do gap: Understanding knowledge application and capacity in long-term care homes. *Social Science & Medicine, 70*(9), 1326–1334.

Bjotveit, A., Wormdahl, E. D., & Tuastad, L. (2016). *Evaluering av prosjektet MOT82. MusikkOppfølgingsTilbud for mennesker med psykisk lidelse i Åsane [Evaluation of the project MOT82. Musical follow-up activities for persons with mental health challenges in the district Åsane in the city of Bergen]*. Bergen: Bjørgvin DPS, Helse Bergen [Bergen Health Trust].

Bolger, L., & Skewes McFerran, K. (2013). Demonstrating sustainability in the practices of music therapists: Reflections from Bangladesh. *Voices: A World Forum for Music Therapy, 13*(2). https://doi.org/10.15845/voices.v13i2.715.

Bronfenbrenner, U. (1979). *The ecology of human development. Experiments by nature and design*. Cambridge, MA: Harvard University Press.

Bronfenbrenner, U. (Ed.). (2005). *Making human beings human: Bioecological perspectives on human development*. Thousand Oaks: Sage Publications.

Bruscia, K. (2014). *Defining music therapy* (3rd ed.). University Park: Barcelona Publishers.

Bunt, L., & Stige, B. (2014). *Music therapy: An art beyond words* (2nd ed.). London: Routledge.

Cole, M. (1996). *Cultural psychology. A once and future discipline*. Cambridge, MA: The Belknap Press of Harvard University Press.

DeNora, T. (2000). *Music in everyday life*. Cambridge: Cambridge University Press.

Flodgren, G., Parmelli, E., Doumt, G., Gattellari, M., O'Brien, M. A., Grimshaw, J., et al. (2011). Local opinion leaders: Effects on professional practice and health care outcomes. *Cochrane Database of Systematic Reviews [Online]*. Retrieved from: http://onlinelibrary.wiley.com/doi/10.1002/14651858.CD000125.pub4/abstract.

Flodgren, G., Rojas-Reyes, M. X., Cole, N., & Foxcroft, D. R. (2012). Effectiveness of organisational infrastructures to promote evidence-based nursing practice. *Cochrane Database of Systematic Reviews [Online]*. Retrieved from: http://onlinelibrary.wiley.com/doi/10.1002/14651858.CD002212.pub2/abstract.

Franks, D. D., & Turner, J. H. (Eds.). (2013). *Handbook of neurosociology*. New York: Springer.

Gold, C., Solli, H. P., Krüger, V., & Lie, S. A. (2009). Doseresponse relationship in music therapy for people with serious mental disorders: Systematic review and metaanalysis. *Clinical Psychology Review, 29*(3), 193–207.

Heine, J. H. (2015). *Cultural psychology* (Third ed.). New York: W.W. Norton & Company.

Helsedirektoratet. (2013). *Nasjonal faglig retningslinje for utredning, behandling og oppfølging av personer med psykotiske lidelser [(Norwegian) National guidelines for assessment, treatment and follow-up of persons with psychotic disorders]*. Oslo: Helsedirektoratet.

Jacobsen, F. (2015). Understanding public elderly care policy in Norway: A narrative analysis of governmental white papers. *Journal of Aging Studies, 34*, 199–205. https://doi.org/10.1016/j.jaging.2015.04.006.

Kleive, M., & Stige, B. (1988). *Med lengting, liv og song [With longing, life, and song.]*. Oslo: Samlaget.

Lov om folkehelseabeid (2011). The Norwegian law on public health work. Retrieved from: https://lovdata.no/dokument/NL/lov/2011-06-24-29

Mekki, T. E. (2015). How do the characteristics of context influence the work of facilitators when implementing a standardised educational intervention targeting nursing home staff to reduce restraint in dementia care? Unpublished PhD thesis. Edinburgh: Queen Margareth University.

Miller, L., Dalli, C., & Urban, M. (Eds.). (2012). *Early childhood grows up: Towards a critical ecology of the profession*. London: Springer.

Mössler, K., Chen, X., Heldal, T. O., & Gold, C. (2011). Music therapy for people with schizophrenia and schizophrenialike disorders. *Cochrane Database of Systematic Review, 2011*(12), CD004025.

Pentland, D., Forsyth, D. K., McCiver, D., Walsh, M., Murray, R., Irvine, L., et al. (2011). Key characteristics of knowledge transfer and exchange in healthcare: Integrative literature review. *Journal of Advanced Nursing, 67*(7), 1408–1425.

POLYFON-nytt. (2017). *News bulletin for POLYFON knowledge cluster for music therapy.*: Retrieved from: http://us10.campaign-archive2.com/?u=fc278c0573acd4be3c76e7f49&id=189c62cb04. Bergen: GAMUT – The Grieg Academy Music Therapy Research Centre.

Rappaport, J. (1981). In praise of paradox: A social policy of empowerment over prevention. *American Journal of Community Psychology, 9*(1), 1–25.

Rolvsjord, R. (2015). What clients do to make music therapy work: A qualitative multiple case study in adult mental health care. *Nordic Journal of Music Therapy, 24*(4), 296–321. https://doi.org/10.1080/08098131.2014.964753.

Small, C. (1998). *Musicking. The meanings of performing and listening*. Hanover: Wesleyan University Press.

Stige, B. (1996). Music, music therapy, and health promotion. In: *Report. International UNESCO-conference, Oslo, September 1995*. Oslo, Norway: The Norwegian National Commission for UNESCO.

Stige, B. (2002). *Culture-centered music therapy*. Gilsum: Barcelona Publishers.

Stige, B. (2006). Toward a notion of participation in music therapy. *Nordic Journal of Music Therapy, 15*(2), 121–138.

Stige, B. (2012). Health musicking: A perspective on music and health as action and performance. In R. MacDonald, G. Kreutz, & L. Mitchell (Eds.), *Music, health and wellbeing* (pp. 183–195). New York: Oxford University Press.

Stige, B., & Aarø, L. E. (2012). *Invitation to community music therapy*. New York: Routledge.

Stige, B., Ansdell, G., Elefant, C., & Pavlicevic, M. (2010). *Where music helps. Community music therapy in action and reflection*. Farnham: Ashgate.

Turner, B. (2004). *The new medical sociology. Social forms of health and illness*. New York: W. W. Norton & Company.

Ward, V., Smith, S., House, A., & Hamer, S. (2012). Exploring knowledge exchange: A useful framework for practice and policy. *Social Science & Medicine, 74*, 297–304.

Wilson, E. O. (1975/2000). *Sociobiology: The new synthesis*. (25th Anniversary Edition). Cambridge, MA: Belknap Press of Harvard University Press.

Wittgenstein, L. (1953/1967). Philosophical investigations. Oxford: Blackwell.

World Health Organization. (1978). The Alma Ata conference on primary health care. *WHO Cronicle, 32*(11), 409–430. See also: http://www.who.int/hpr/NPH/docs/declaration_almaata.pdf.

World Health Organization. (1986). *Ottawa charter for health promotion*. Geneva: World Health Organization.
World Health Organization. (1998). *Health promotion glossary*. Geneva: World Health Organization.
World Health Organization. (2012). *Governance for health in the 21st century*. Copenhagen: World Health Organization, Regional Office for Europe.

Chapter 9
Music *as* Participation! Exploring Music's Potential to Avoid Isolation and Promote Health

Karette Stensæth

Introduction

Together, social isolation and loneliness are about to become the biggest public health threat of our age (Cacioppo and Hawkley 2003; House et al. 1998), especially for populations where socially isolating preconditions are a risk (Berg 2009). There is a great need to explore new ways to fight social isolation and to find meaningful ways for people to be with others while engaging in participation, both on an individual level and on a community level. Music, especially through the development of the discipline of music therapy, offers novel ways of approaching some of the challenges connected to social isolation.

This essay suggests the field of public health to look into the value of the music therapy knowledge and to consider novel measures for health-promoting participation for all. The essay is structured as follows: The first part (1) includes reflections upon our understanding of social isolation, loneliness and participation. This part is presented as an introductory background for the first question: What are the potential connections between isolation and musical participation? The next part (2) is called "music *as* participation" and reflects upon research projects in the Norwegian child welfare that involve the use of music interventions, such as music and drama, songwriting and performances, as a way to promote participation among children and youth. Most of these projects involve a professional music therapist or other music and health workers. This second part responds to the question: Can music activities become a resource for avoiding isolation and promoting participation instead?

K. Stensæth (✉)
Center for Research in Music and Health, the Norwegian Academy of Music, Oslo, Norway
e-mail: Karette.A.Stensaeth@nmh.no

L. O. Bonde, T. Theorell (eds.), *Music and Public Health*,
https://doi.org/10.1007/978-3-319-76240-1_9

Part 1: Social Isolation, Loneliness and Participation

Traditionally, the field of public health is about promoting and protecting the health of people and the communities where they live, learn, work and play. The aim is to prevent people from getting sick, suffering or injured and to promote wellness and our quality of life, by encouraging healthy behaviours. Public health is also about spreading the word about ways to stay healthy and giving science-based solutions to problems.[1]

Social isolation, loneliness and participation are vital concerns in the public health area. These topics have resulted in a large amount of studies. A quick search on Google Scholar on "social isolation, loneliness, participation and public health research" gives more than 630,000 hits. Many interventions have been tried to break the cycle of isolation, and most of them conclude the need to recognize and adapt to not only the community but also individual needs. The research includes studies from many disciplines, especially social psychology. In a special section of the March 2015 issue of the journal *Perspectives on Psychological Science*, psychologists took stock of some of the potential causes and risks of loneliness, as well as possible treatments. *Huffington Post* (HP) refers to this research that estimates that one in five Americans suffers from persistent loneliness. Sbarra (2015), who is one of the researchers that HP refers to, says that the tendency is the same elsewhere[2]:

> Many people thought of loneliness as a transient state – something most everyone experiences but that is relatively short-lived. As we learned that some people are chronically lonely, we began to see that the topic has considerable public health importance. (Loc. cit.)

Isolation

Loneliness and social isolation are not the same phenomena. Social isolation is characterized by an absence of social interactions, social support structures and engagement with wider community activities or structures (Holt-Lundstad et al. 2015). Loneliness refers to an individual's personal, subjective sense of lacking connection and contact with social interactions to the extent that they are wanted or needed (ibid.). There is also a third aspect called social inaction, which describes a state where individuals choose or are unable to take part in in social action and of various reasons are disconnected from concepts of "we-ness" and civic society.[3]

Social isolation and loneliness are however often connected and are therefore talked about as the same type of public health problem. Lonely people are often

---

[1] Retrieved 21 April, 2017: https://www.apha.org/what-is-public-health.

[2] Retrieved 20 April, 2017: http://www.huffingtonpost.com/2015/03/21/science-loneliness_n_6864066.html.

[3] Retrieved 23 April, 2017: https://publichealthmatters.blog.gov.uk/2015/12/08/loneliness-and-isolation-social-relationships-are-key-to-good-health/.

socially isolated and vice versa. Psychologists (Holt-Lundstad et al. 2015) discovered in a meta-analysis that loneliness and social isolation better predicted premature death among populations aged less than 65 years, despite older people being more likely to be lonely and having a higher mortality risk overall. Their review of studies indicated that loneliness increased mortality risk by 26%. "The effect of this is comparable to obesity, something that public health takes very seriously", said lead author Julianne Holt-Lundstad, "we need to start taking our social relationships more seriously" (ibid., no paging). The BBC, in their program *The Age of Loneliness*, builds on this research and says social isolation is becoming "epidemic" among young adults, and even more common among the elderly, with some reporting that the numbers of those feeling lonely and isolated have doubled since 1980.[4]

Independent factors causing social isolation and loneliness have been associated with lower reported life satisfaction, alcoholism, suicide and physical illness (House et al. 1998). More recent research has suggested the correlation with morbidity and mortality is stronger for social isolation than loneliness (House 2001).

All in all, modern life seems to make us all lonelier, a trend that is about to become the next biggest public health issue on par with obesity and substance abuse. The report from Public Health England from 2015, for example, says that social isolation and loneliness together are one of the biggest threats to public health of today, twice as deadly as obesity and as potent a cause of early death as smoking 15 cigarettes a day. Psychological, behavioural and biological challenges caused by social isolation and loneliness are one of the biggest threats to public health in the western society. Yet, compared to health behaviours such as smoking and obesity, much less is known about how and why social isolation affects health (Cacioppo and Hawkley 2003; House et al. 1998; Uchino et al. 1996).

Cornwell and Waite (2009) consolidated multiple measures of social isolation. They found that several authors have previously identified central components of isolation. Some distinguish between social loneliness, as the lack of integration and companionship, and emotional loneliness, as the lack of an attachment figure. Other authors similarly contrast isolation (as the opposite of integration) with loneliness (as the opposite of embeddedness).

Following these distinctions, and building from the disciplinary approaches of sociology and psychology, Cornwell and Waite (2009) suggest two forms of social isolation: social disconnectedness and perceived isolation. In their study (ibid.), they found results that indicate that social disconnectedness and perceived isolation are independently associated with lower levels of self-rated physical health. They suggest that the association between disconnectedness and mental health may however operate through the strong relationship between perceived isolation and mental health. They therefore conclude that health researchers need to consider social disconnectedness and perceived isolation simultaneously. However, it cannot be assumed that social isolation is a universal problem. Although highly correlated, social isolation does not necessarily go hand in hand with loneliness. Solitude may sometimes be desired by many individuals (Henderson 2015).

[4] The BBC program is still available at: http://www.bbc.co.uk/programmes/b06vkhr5.

In a study, Cacioppo and Hawkley (2003) analysed data about three main types of treatment for loneliness: group therapy, individual treatments (working with a therapist to improve befriending skills or to minimize negative beliefs that might contribute to loneliness) and community interventions (events focused on reaching out to lonely people). Examining a body of existing literature on the subject, the researchers concluded that the most promising line of treatment for loneliness is individual therapy that addresses the thought patterns and beliefs, such as low self-esteem or shame, that prevent a person from connecting with others. Despite not mentioning music therapy in their study, these results point to important aspects that music therapists can take into consideration when working with people dealing with loneliness.

Participation

The opposite of isolation is sometimes described as participation. In the everyday usage of the word, participation labels our interest in taking part in something. It might derive from the Latin *participare*, to share, impart and partake of, but it might also derive from the Latin *partem carpere* – that is, specifically to take something from someone (Myetmyology.com 2008 in Stensæth et al. 2014). The latter derivation connotes a certain dimension of power, which might explain the political applications of participation, as a motivating force for democracy, for example. As a noun, participation points to the act of sharing in the activities of a group and/or the condition of having something in common with others (as fellows, partners, etc.) (ibid.).

Normally, participation assigns several aspects: people's preferences and interests; what they do, where, and with whom; and how much enjoyment and satisfaction they find. For participation to be meaningful, it is crucial that people have a sense of choice or control over their activity. They also need a supportive environment to facilitate their attention, a focus on the task rather than the long-term consequences, a sense of challenge from the activity and a sense of mastery over it. Therapists may refer to this as *the just right challenge*.

Participation creates a complex area in public health and has become a central construct in health care, rehabilitation and various forms of therapy (including music therapy). In these areas, participation is often referred to as a means of describing involvement in various life areas (Berg 2009; Imrie and Hall 2001; Law 2002).

The notion of participation pervades many fields and research. The ideal behind universal design is to include broad-spectrum ideas meant to produce buildings, products and environments that are inherently accessible to all people, with and without disabilities (Imrie and Hall 2001). In research, the ideal of participatory research, which is about engaging workers and other stakeholders in systems development to enabling them to serve as co-designers, is much used (see Stensæth et al. 2014).

In order to promote more participation of citizens with disabilities in society, the WHO (2001) changed their view on health from considering it a "consequence of disease" to considering it a "component of health" (p. 4). This integrated understanding of health fuelled a social model of participation, one that included environmental factors from the individual's most immediate environment to the larger communal environment, encompassing social and institutional structures (ibid; see also Stensæth et al. 2014). The latter model on participation, which draws a social health model, is, as we shall see in the following, adequate for the perspective in this essay.

Also, typical of our times is that technology has made it possible for us to be more connected than ever before. With its change from analogue (device-determined) to digital (program-determined) technology, our ways of living have changed touching all areas of participation in our everyday life. Technological aids and our digital devices have become immediate means for connecting with other people. Frequently, technology is presented as the (economic) solution to the challenge of loneliness and social isolation. A meta-analysis conducted by Choi et al. (2012) found that computer and Internet training interventions were significantly effective in decreasing loneliness among the older populations but that there were concerns about sustainability and ease of use.

The social media do not replace our need for human contact and participation in meaningful activities with others and might actually exacerbate the problem making us feeling lonelier and more isolated (cf. Turkle 2011). Despite loneliness being often associated with older people, new research shows that 6 in 10 teenagers are sometimes lonely and 1 in 20 never spend time with friends outside of school (Lau 2016). Lau's report (ibid.) found that increased online interaction does not damage teenagers' social skills. The findings showed a small relationship in the opposite direction: teenagers with better ability to form friendships outside school reported more online usage thus suggesting that online usage could support the development of their social skills. Lau argues however that "social intelligence", defined as the ability to apply our understanding of people's emotions to decide the appropriate form of interaction with others, will become increasingly important to future generations. While important as a means of practicing social skills, online interaction is not understood as a substitute for real-life interaction. Not only is online interaction associated with more loneliness in later life, this form of communication alone is not adequate in preparing young people for the challenges of the workplace and other everyday life forms (ibid.).

Summing Up Part 1

To sum up this part, we see that for many lonely people, reducing social isolation and helping them link up with others often reduce loneliness. The digital revolution does not change the fact that we are feeling lonelier. Increasing opportunities to make friends on a structural level do not always result in reducing a person's loneliness. While feelings of loneliness and social isolation lead to mental health problems such as depression, anxiety and a whole assortment of ill-health effects, simple face-to-face human interaction is one of the best and easiest ways to increase human happiness and trusting human relationships (Turkle 2011). Similarly, people who feel they have supportive and reliable communities are less likely to feel lonely (Cornwell and Waite 2009; Turkle 2011). Therefore, participation, to become socially valuable for the individuals and their communities, require active part taking. Literally speaking, people need to feel that they can *take* part, both on their own and together with others.

Part 2: Music *as* Participation

To promote participation through music and musical activities is one of many novel approaches on the use of creative and aesthetical means to fight isolation (MacDonald 2012; Bonde 2009).[5] In the large picture, however, there is still little research on the connections between music and participation with the attempt to reduce isolation. The interest in the positive effects and values on the connections between music and participation has however increased enormously over the last two decades (cf. Bonde and Trondalen 2012; MacDonald 2012, 2013). It is the promising value/results from this research that this essay describes in this part.

Recent research in sociology and psychology of music has challenged the traditional idea of music as an autonomous art form. Researchers have identified functions of music that exert a regulatory influence upon both mind and body that contribute to well-being and improved health and life quality (Bonde 2001; DeNora 2000; Ruud 2016). Other research has demonstrated that musical practices represent a social resource – a means of bonding and connecting (Bonde 2011; Malloch and Trevarthen 2009). At the same time, research on everyday musical practices and music making in amateur activities has shown how music is valued as a social resource and a means of social bonding and connecting, as well as a means of negotiating identity (Ansdell 2015; Ruud 1997; Stensæth and Næss 2013; Stige and Aarø 2012; Trondalen 2016b).

There is a growing cultural tendency to study music activities and music making as a communication arena with health prospects for everyone, both young and old (for various overviews, see Ansdell and Pavlicevic 2009; Bonde 2009, 2013; Bonde and Trondalen 2012; MacDonald 2012, 2013; Ruud 2016). This research shows that a healthy participation, both on an individual level and on a community level, can be studied in the light of a musical and aesthetic practice (Aldridge 2004). MacDonald (2013) suggests a model to show how various musical fields relate to one another in the perspective of music, health and well-being. He lists up four different musical arenas as the major ones: music education, everyday use of music, community music and music therapy (the latter with another smaller area of music medicine attached to it). All of these areas, although they possess different levels of

[5] In this part, the essay refers to *isolation* as a notion that contains all those aspects described above with regard to loneliness and social isolation.

expertise, share an interest in their understanding of musical participation as a public health means.

On a community level, there is a growing tendency to view music as a source for successful interventions for participation. Many symphony orchestras implement the so-called social projects into their outreach programs, where musicians perform concerts outside the large concert halls to approach the public where they are (on the street, in the tube stations, etc.) and/or invite people to play with them who never would have had the chance to do so otherwise. Professional musicians – sometimes together with music therapists – also perform music in hospitals and nursing homes.

On the individual level, music is revealed as an instrument in the repertoire of "self-technologies", which aims at regulating our bodies, emotions and cognitive orientations (DeNora 2000; Ansdell 2015). Research describes also how people use technology as a way to cope in their everyday lives (Beckmann 2014; Batt-Rawden et al. 2005; Skånland 2013). Some studies show that the music-listening practices among youngsters, who are often the greatest music consumers, become health promoting when it helps them cope and regulate their feelings (Beckmann 2014; Skarpeid 2009; Stene 2009; Skånland 2013). The youngsters even identify themselves through their music and talk of it as if their music were a friend who is always there for them. They experience music as a resource with which they can express feelings, explore identity, socialize with peers and/or regulate their emotions as well as their distance to adults (Baltazar and Saarikallio 2016; Hense and McFerran 2017; McFerran 2010, 2016; Laiho 2004; Saarikallio 2016; Saarikallio et al. 2017).

Music Therapy and Participation

Music therapy has over the last 50 (or more) years developed several meaningful practices and activities for populations who are at risk of becoming isolated. As a discipline and an expert field, music therapy has also developed useful research and valuable theory, especially in the last 15–20 years. More and more, we see that music therapy theory serves as a backdrop for a common understanding on music as a source of health-promoting participation in other fields. Today we hear people talk like music therapists about how music might strengthen our self-esteem and feeling of mastery, health *and* participation (cf. Ruud 2016; Bonde 2009).

In music therapy the benefits emerging from participating in music activities are assessed by their degree of health value. Ruud (2010, 2014) argues that in music therapy, health, more than a biomedical state, is highlighted as a personal experience that is understood as a component of health. In this perspective, health, rather than a fixed state, is regarded as a fluid state that can be influenced, for example, through meaningful musical participation and the experiences deriving from it (ibid; Stensæth and Næss 2013). This could include many forms of music activities, such as music listening, singing in a choir or playing in a band and many more (cf. Ruud 2010, 2014; Bonde 2009). Ruud (2010, 2014) calls for such an experiential focus on health where music becomes a way to mobilize oneself together with others towards a better quality of life. Health in this picture is equivalent with a feeling of well-being and the capacity and capability for participation and action (or, in the case of poor health, as a state of suffering or a lack of ability to act, which easily leads to social isolation) (ibid; Stensæth and Jenssen 2016).

Music, and especially active music making, often proves that there is a strong conceptual connection between the state of well-being and the ability to act and to take part (Nordenfelt 1991). Active music making, in this sense, almost *becomes* participation. Small's (1998) notion of "musicking" is particularly evocative in this music therapy perspective precisely because it emphasizes music as action and as a doing that could enable participation. For Small, the act of musicking establishes, in the place where it is happening, a set of relationships, and it is in those relationships that the meaning of the act lies. Musicking becomes this way an active means of relating to – and participating in – the rest of the world, together with others.

Ansdell and Pavlicevic (2009) suggest the notion "collaborative musicing" to cover group music activities that explain the connection between our inherent musicality (nature) and our social musical practices (culture): "Collaborative musicing builds community through making music together" (ibid, p. 369).[6] In their model, social experience and musical experience activate the functions of (1) human communication (in the understanding of sharing and being together) and (2) human collaboration. By fusing models for musical development and social development, described as the process moving from the "I" through the "We" and then to the "Us", they show how musicking is linked to social development. According to Ansdell and Pavlicevic (2009), musical development builds on the universal human capacity, an inborn communicative musicality (cf. Malloch and Trevarthen 2009). This universal human capacity, together with certain cultural contexts, is basic for the individual's ability to use the capacity (musicianship). The musicianship in turn leads to active participation in music activities (musicking). They explain that our musicality is more intimate on the I-level and broadest with more people (and communities) on the Us-level. An individual's – or a population's – musicality might however be damaged. To "repair" it is difficult; it depends on the type of damage and the cultural possibilities to engage the individual in the community. Sometimes musicking in large groups is useful for the repairing. Sometimes, musicking to repair the more basic experiences of "I and You" is needed (ibid.).

The relationship between music and young people has always been strong, sometimes expressed by themselves as a question of "life and death" (Beckmann 2014). Krüger and Stige (2015) think it is time to claim that music is a right and that the need for systematic music therapy work in the child welfare could be based on a human rights perspective (cf. Curtis and Vaillancourt 2012). Norway has ratified the child welfare convention, which implies a commitment to follow the value-related foundations the convention is built upon. Christiansen et al. (2015) think that the convention clarifies the young people's right to participation, but that there is a lack of adequate practices to ensure children and young people's right to par-

[6] Ansdell and Pavlicevic write musicking without the k, which is what the music educator David Elliot did in his early writings too.

ticipate and collaborate (cf. Blaustein and Kinniburgh 2015; Curtis and Vaillancourt 2012). The young people's experience of being isolated has many explanations, and some of them need to be mentioned before returning to the role of the music as participation.

Children and Adolescents Within the Child Welfare System

Nearly 50,000 children and adolescents received services from the child welfare system in Norway in 2010. About 2000 of these individuals were living in child welfare institutions. This number represents an increase of 7.1% from 2009 to 2011.[7] In 2015, the number of 53,439 children and youth from 0 to 22 years received help. Fourty percent of these were placed in foster care.[8] The health care offers services on more than twice as many occasions as there are recipients. At the end of 2015, healthcare services were given on 81,206 occasions to help 36,811 children and adolescents. This shows that children and families often have complex needs simultaneously and that they receive different types of services from the child welfare.

The dream of children and adolescents within the child welfare system are to live a so-called normal life: to wake up every morning with their family, eat breakfast, go to school, be with friends, attend leisure activities, etc. (Storø 2016). However, this is often not a simple task for them.They have not lived a stable life in a family with trusting significant others, and many of them suffer from development trauma. As a term, development trauma describes the severe traumatization caused by persistent violations in the earliest childhood, often in situations where the violations are inflicted by someone the children feel close to, like father, mother, step parents and grandparents. These violations might be of a physical or a mental character (or both). Prof. van der Kolk (2017) and other professionals at the Trauma Center describe childhood trauma as the single most important public health challenge.[9] At the core of traumatic stress is a breakdown in the capacity to regulate internal states. Unfortunately, all too often, medications are used to deal with and master their uncomfortable physical sensations. On this background, we understand that children and adolescents in the child welfare system are vulnerable and that they easily develop a feeling of social disconnectedness and perceived isolation (cf. Cacioppo and Hawkley 2003; Lau 2016). Not feeling that they belong anywhere easily creates problems with identity (cf. Ruud 1997), low self-esteem and/or shame. It also explains why children and adolescents in the child welfare system often end up quitting school, why few of them get normal jobs and too many have troubles with drugs and addiction.

---

[7] Barnevernpanelet, 2011, retrieved 23 April, 2017, see: www.regjeringen.no/globalassets/upload/BLD/Barnevern/2011/barnevernpanelets_rapport.pdf).

[8] Retrieved 23 April, 2017, see: www.bufdir.no/Statistikk_og_analyse/Barnevern/.

[9] Retrieved May 9, 2017 at http://www.traumacenter.org/products/pdf_files/preprint_dev_trauma_disorder.pdf.

Their challenges have however the potential to be largely resolved by appropriate prevention and intervention (Blaustein and Kinniburgh 2015). Safety, predictability and "fun" is essential; most of all, they need activities that give them a sense of pleasure and mastery and help them find new ways of coping (ibid.). Activities that give them a "break" from problematic inner states are desirable, and non-medical treatment is very much in demand.

Music Therapy and Participation in the Child Welfare

Music therapy is meant to provide a meaningful and complementary treatment in mental health. The Norwegian Directorate of Health (Helsedirektoratet 2013) underlines the need for the treatment to be performed by therapists with approved education, and it should be included in mental health treatment within the community. From what we know so far, and according to the latest music therapy research and the Directorate, there are reasons to believe that music activities and music therapy provides adequate participation practices for children and adolescents within the child welfare. Expressing oneself through music not only affords a form the young people find familiar and motivating; it also creates resources for agency, performing and sharing, while providing "non-threatening" meetings with professionals (Stensæth et al. 2016).

There seems to be a lack of adequate practices, in which the young people are motivated to engage and can ensure their right to participate and collaborate. Music therapy seems therefore in many ways to be a meaningful approach for them – one that provides motivating ways for them to actively *take* part – and thus seems to be a good alternative to traditional therapy (i.e. talking), especially if the youngsters have little trust in words. Music therapy projects in the child welfare show how the use of music therapy might give adolescents new possibilities for participation and experiences of community, as well as space for the development of their own personal resources (Krüger and Strandbu 2015; Stensæth et al. 2016). However, one way of participating (i.e. music) should not exclude another; rather they could complement each other (Strandbu 2011).

An objective behind community music therapy (Stige and Aarø 2012), which sees music therapy in a larger context, is exactly to establish music-related communities for inclusion, for example for unaccompanied refugee minors (Hunt 2005; Enge 2015; Roaldsnes 2016). Another purpose is to foster young people's participation and autonomy in transitions between living in residential care and moving to their own places (Krüger et al. 2014).

The fresh anthology, *In transit – between to and fro* (Stensæth et al. 2016), is a collection of projects in the child welfare that show how music is used *as* participation.[10] The book's title is taken from a line in the song called *Transit,* which was written by a girl called "Ida", who participated in the music theatre project

[10] The title is translated from Norwegian by the present author.

called *Come closer* (Kom nærmere) in Bergen, Norway.[11] "In transit" is in many ways a suitable metaphor of the feeling described by Ida and many other children in the child welfare system: the feeling of living somewhere in the transitions, in between to and fro, either geographically, psychologically or emotionally. Understandably, not feeling that you belong anywhere and to continuously live your life on hold, never knowing where to go and when to move on, who to connect to or what comes next, is a stressful life. Symbolically Ida sings, "That is why I always keep my jacket on."

The book *In transit – between to and fro* refers to several qualitatively oriented Norwegian music therapy projects in child welfare that provide us with in-depth descriptions of the benefits of the participants' musical participation:

- In some of the projects, music is used to regulate and alleviate feelings, which in turn help the young population to experience stability in their life (Helle-Valle 2016; Trondalen 2016a, b).
- Other projects show that when facilitated professionally and with care, participation in the music activities give the children and youth experiences of success, joy, hope and recognition (Fuglestad 2016; Helle-Valle 2016; Kristiansen 2016; Mandal and Bergset 2016; Roaldsnes 2016; Strandbu et al. 2016; Storø 2016; Trondalen 2016b; Tuastad 2016).
- Roaldsnes (2016) in her interviews of unaccompanied refugee minors in the child welfare in Norway found that the music groups with them were important exactly because they created valuable positive emotions that gave them much needed distraction from their traumas.

Many different methods are used in the various music approaches in the music therapy projects presented above: songwriting, song sharing, (systematic) music listening, musical interplay, band playing, improvisation, performance, etc. The aim is to create ways for the young people to express themselves, so that the music becomes a resource that helps them create their own individual autobiographical narratives (Krüger et al. 2016; Stensæth et al. 2016, see also Krüger and Stige 2015).

Common for the projects is a focus on music as a means for health-promoting participation (Stensæth and Jenssen 2016). The projects challenge the problem-oriented focus on the child welfare with a resource-oriented focus, which emphasizes music as an empowering and collaborative means (Ruud 2010). This type of resource-oriented focus is typical for the humanistic perspective on music therapy with a view on the individual as a unique biological, psychological and social being (ibid.). This view, which has become traditional in Norwegian music therapy, and is a perspective that suggests that music activities create vital tools to developing mastery and self-confidence, is needed among children and adolescents within the child welfare (cf. Blaustein and Kinniburgh 2015).

---

[11] Read more about *Come closer* on their website: https://www.aleris.no/Her-finnes-vi/Region-Vest/Region-vest/Teaterlag/ . Ida wrote her song together with music therapist, Viggo Krüger and theatre instructor Morten Lorentzen in *Come closer* and it is part of the music intervention. The song, *Transit*, can be heard on YouTube and Spotify.

The projects are stories – as is Ida's story – showing that *taking* part through music emerges as an existential value and a social potential where the individuals can flourish through musical expression. They also show that music is helpful in building communicative musicianship where the young population can express and share their feelings in the here and now *and* experience recognition from others. *Through* music, they can speak up and we can listen to their voices, literally speaking. Also, because music offers alternative (sometimes non-verbal) forms of communication, the music therapists have a means with which they can execute professional expertise to offer adequate help for the ones who have been exposed to traumatic events and/or have difficulties with verbal expression.

The Use of Music for Young People to Move Closer to Oneself, to Others and Society

In *Come closer,* the project in which Ida participated, both youngsters and leaders participated in songwriting, drama, band playing, improvisation, concerts and music theatre performance or help out with sound, technology or other tasks. The participants can also talk with the competent leaders about their experiences and needs. On the Come closer website, the youngsters describe that they use music to show (the rest of the world) what experiences they have from living within the child welfare. The musical participation affords a channel for them to speak up.

Come closer is an example of important aspects of the role of the music: it gives the children and adolescents a personal voice and a voice that becomes audible for the public through their concerts and their CDs. Music is not given one single meaning; its value is linked to how it is used and is related to contextual factors such as culture, knowledge or communication (ibid; Krüger and Stige 2015). Music (literally) affords a way of children and adolescents to take part on their own premises (cf. Backe-Hansen 2016). Music activities in fact become a tool for them to build constructive social relationships with other children and youth.

The music therapist and his co-workers in *Come closer* describe that the idea is to give young people within the health-care system an opportunity to participate in meaningful cultural activities that also provide experiences of mastery, community and belonging (Strandbu et al. 2016).

Children and adolescents who have participated in other music therapy programs within the child welfare report that the music therapy gives them "space" and a chance "to breathe" (Krüger 2016; Krüger and Stige 2015; Strandbu et al. 2016). They also tell that it helps them create positive associations and a better self-image (ibid.). Importantly, they describe that taking part in music activities provides unique opportunities for them to be with adults and peers, which is something they miss the most (Krüger and Stige 2015).

Backe-Hansen (2016) points out that recently child research in the social sciences has moved from being research *on* children to being research *with* children. Today we no longer see children as an object of knowledge acquisition but as acting subjects who have their own voice. They in fact now have a right to speak up and to take part on their own terms, and we (the grown-ups around them) are committed to listen to them before making decisions (i.e. Barnevernloven § 4–1).[12] Backe-Hansen (2016) argues that participation in music activities, such as the ones in *Come closer*, could offer adequate education for young people in democracy and citizenship. Participation, preferably in groups, is something that this population needs to practise, she says, and importantly, this practice should be offered in a form that holds the trust and respect in a protective relationship with significant others. Because adolescents and adults can find shared interest in music, they can also work with issues like power relations and lack of communication, which can often occur in the context of the social work and with the population in question (ibid.).

The present author thinks that to build healthy communities like this could be a small start that in the long run could be of vital ecological importance for the society and our public health. The opposite, and especially the extreme opposite, is dangerous and scary. Khan, a British reporter, in her documentary film of young Jihad fighters, says that their radicalization is primarily explained by the pain the young people feel by meeting racism, exclusion, marginalization, loneliness and social isolation.[13] *Healthy* participation is therefore basic, and music, when facilitated systematically and professionally by a music therapist, should perhaps be considered a structuring resource for a complex set of participatory practices (Krüger and Stige 2015; Strandbu et. al. 2016).

Dialogical Aspects in Musical Participation

In the music therapy projects referred to in this essay, dialogue is highlighted as a key element in musical participation (Stensæth and Jenssen 2016). Dialogue involves subjectivity and starts with the face-to-face position. It is about the basic recognition of the other even if the other means something else, which is highly important for the development of the basic "social intelligence", which Lau (2016) called for. However, because Ida's transit situation makes it difficult for her to trust other people (cf. Backe-Hansen 2016), she needs professionals who are (both personally and professionally) able to position themselves as *close* others for her. A close other is one who listens openly and relates seriously and with sympathy, attachment and compassion to all possible communicative signals from another individual. This is the first and basic step in the process towards social participation. Normally, this is not something we think of, but for children and adolescents in the child welfare with trauma history, this part is a vital source for their participation.

[12] Retrieved 6 May 2017 from https://www.google.no/search?q=Barnevernloven+%C2%A7+4-1&ie=utf-8&oe=utf-8&client=firefox-b-ab&gfe_rd=cr&ei=MyoPWfv1EOPk8Ae0kouAAQ.

[13] See the film at: https://tv.nrk.no/program/K;TE30000614/jihad-hellige-krigere. Stensæth and Jenssen (2016) reflect upon this too.

The close other is basic for building trust and creating empowering micro-dialogues. The role of a close other assumes a certain mind-set and an ethical awareness of the value of the face-to-face positioning. It could also be seen as a type of world-view that unifies with Mikhail Bakhtin's dialogue philosophy (Bakhtin 1981), which emphasizes that we are all born with a dialogical mind that is reflected by other people's minds. To relate to another in a dialogical sense calls for an ethical and aesthetical awareness in the I-You relation. Returning to Ida and her lifeworld, this means that she needs a closer other to become a true You, who hears, sees, and loves her as the other. This view contradicts the cognitive awareness where the other becomes (solely) an object for participation or learning, therapy and research (cf. Backe-Hansen 2016).

This author thinks that music *as* participation becomes dialogic when the I becomes competent within a We-community (cf. Ansdell and Pavlicevic 2009 and their model on collaborative musicking), when there is a mutual acceptance and a willingness among the participants to come into play with the other(s). As an experienced music therapist, this author has seen how valuable music can be as a dialogic means. Perhaps music is so useful as a dialogic means because it helps people to feel; it helps people to feel when they are numbed, such as by traumas and tragedies. Recognizing the feeling could be the first step out of any problem. After that follows the urge to express and share the feelings. That is perhaps music's most important role: to offer people a voice for expression, we need to have meaningful and relevant activities at hand, so that the individuals can actively take part.

Summary and Critical Remarks

This essay has portrayed music *as* participation, as a resource or form of social capital that people can use to build social networks and provide meaning and coherence in life with a prospect of avoiding isolation.

The essay suggests that music is a powerful (yet still a somewhat hidden) means for participation that has the potential to hinder social isolation. Participation in music activities, like the music therapy projects referred to in this essay, involves healthy musicking, which incorporates the participants' desire *to do* (action) *something* (activities) *meaningful* (intentional) together (*intersubjective and interpersonal*). Health in this regard is a quality of human participation and coexistence engagement (cf. Hallstead 2013) and a health-performing practice (Stensæth and Jenssen 2016, see also Chap. 8 by Stige).

The essay refers to music therapy projects that show that it is evident that music as participation requires a certain experience of dialogical *sharing*. This sharing involves collaborative music(k)ing (cf. Ansdell and Pavlicevic 2009) and a type of togetherness that is built on a common focus in the doing. The face-to-face-position and an empathic recognition of the other as a subject are basic in these human musical participation processes; it allows perhaps the music to create a simple but a very basic aspect to practise and build social intelligence and the ability to connect with others (cf. Lau 2016).

There are of course several challenges connected to the combination of music and participation that this essay has not touched upon. Given music's role and function in the lives of young people in today's digital society, the discussion regarding the excessive availability of music includes the question whether our society always exploits music as health-promoting participation. Along with Krüger and Stige (2015), this essay questions if we, both music therapists and other music workers, have sufficient knowledge concerning the social roles and possibilities for citizen participation that all categories of the use of music may provide? Do we have enough knowledge concerning the technologies of music or pop culture, for example? The potentials of participation in the combination of technology and music require certainly careful attention in the future, and there is a need for collaborative research projects on the benefits and dangers of the use of music as participation in public health. This research should incorporate multidisciplinary collaboration across all disciplines, including music therapy.

However, instead of generalizing, we need to contextualize the values deriving from musical participation. There is also a need to take into account that music is a strong medium that could be used as a means to engage people in unhealthy participation too (Stensæth and Jenssen 2016), particularly for vulnerable young people who might use music to engage with emotions that do not sufficiently account for their current state of well-being (McFerran 2016). We know that music can sometimes strengthen psychological distress (McFerran 2010).

People with addiction problems, because they tend to associate certain music to drugs and the excitement coming from it, use, for example, music as a way to legitimize their abuse (Kristiansen 2016). Such challenges prove that music therapy research is needed to provide a professional and systematic control on the development of both practice, theory and research in the use of music to empower participation and avoid isolation. The expertise from music therapy is especially salient working with vulnerable populations. For Ida to participate, the qualifications of her collaborators as trustworthy and empathic close others became just as important as the musical qualifications.

Music therapy, with its various active music-making practices, is unfortunately still not an integrated part of the Norwegian child welfare practice. The picture is slowly changing, however, and recently, in January 2017, the Bergen health enterprise in Norway announced that music therapy is mandatory for young people with mental health problems. This means that health clinics in Bergen with mental health-care services have decided upon a strategy trying to hire a professional music therapist. This is promising, and hopefully we will see similar strategies in other departments with other populations, not just in Bergen and Norway but all over the world (see Chap. 8 by Stige).

This essay shows that music, and music therapy in particular, when it is recourse oriented and used with care, might offer a radical perspective to the society's isolation challenges. The theories and the projects referred to in this essay prove that music could be practised as an anti-authoritative communication event, one that represents a non-medical approach of creating and cultivating local democratic micro-cultures. We need however more research to prove *how* and *why* healthy musicking is valuable and an effective means for participation, which also has the potential to hinder social isolation – for all.

References

Aldridge, D. (2004). *Health, the individual and integrated medicine. Revisiting as aesthetic of health care*. London: Jessica Kingsley Publishers.

Ansdell, G. (2015). *How music helps in music therapy and everyday life*. Surrey: Ashgate.

Ansdell, G., & Pavlicevic, M. (2009). *Community music therapy*. London: Jessica Kingsley Publishers.

Backe-Hansen, E. (2016). Barn og unges medvirkning i barnevernet – hvorfor og hvordan? In K. Stensæth, V. Krüger, & S. Fuglestad (Eds.), *I transitt –mellom til og fra. Om musikk og deltagelse i barnevern. Series from the Centre for music and health* (Vol. 9, pp. 37–48). Oslo: Norges musikkhøgskole.

Bakhtin, M. M. (1981). *The dialogic imagination*. Austin: University of Texas Press.

Baltazar, M., & Saarikallio, S. (2016). Toward a better understanding and conceptualization of affect self-regulation through music: A critical, integrative literature review. *Psychology of Music, 44*(6), 1500–1521. https://doi.org/10.1177/0305735616663313.

Batt-Rawden, K. B., DeNora, T., & Ruud, E. (2005). Music listening and empowerment in health promotion: A study of the role and significance of music in everyday life of the long-term ill. *Nordic Journal of Music Therapy, 14*(2), 120–136.

Beckmann, H. B. (2014). Den livsviktige musikk. En kvalitativ undersøkelse om musikk, ungdom og helse. PhD-thesis. NMH-publikasjoner 2014:9. Oslo: Norwegian Academy of Music.

Berg, M. (2009). Hva er deltagelsefor barn som har funksjonshemming? *Ergoterapeuten, 1*(09), 1–5.

Blaustein, M. E., & Kinniburgh, K. M. (2015). When age doesn't match stage: Challenges and considerations in services for transition-age youth with histories of developmental trauma. Regional Research Institute for Human Services, Portland State University. *Focal Point: Youth, Young Adults, & Mental Health. Trauma-Informed Care, 2015*(2), 17–20.

Bonde, L. O. (2009). *Musik og menneske: introduktion til musikpsykologi*. Roskilde: Samfundslitteratur.

Bonde, L. O. (2011). Health musicing—Music therapy or music and health? A model, empirical examples and personal reflections. *Music and Arts in Action, 3*(2), 120–138.

Bonde, L. O., & Trondalen, G. (2012). Music therapy: Models and interventions. In R. A. R. Mac Donald, G. Kreutz, & L. A. Mitchell (Eds.), *Music, health and wellbeing* (pp. 40–61). Oxford: Oxford University Press.

Cacioppo, J. T., & Hawkley, L. C. (2003). Social isolation and health, with an emphasis on underlying mechanisms. *Perspectives in Biology and Medicine, 46*, 39–52.

Choi, M., Kong, S., & Jung, D. (2012). Computer and Internet interventions for loneliness and depression in older adults: A meta-analysis. *Healthcare Informatics Research, 18*(3). https://doi.org/10.4258/hir.2012.18.3.191.

Christiansen, Ø., Bakketeig, E., Skilbred, D., Madsen, C., Skaale Havnen, K. J., Aarland, K., et al. (2015). *Forskningskunnskap om barnevernets hjelpetiltak*, Uni Research Helse, Regionalt kunnskapssenter for barn og unge (RKBU Vest).

Cornwell, E. Y., & Waite, L. J. (2009). Social disconnectedness, perceived isolation and health among older adults. *Journal of Health and Social Behavior, 50*(1), 31–48.

Curtis, S., & Vaillancourt, G. (2012). The children's right to music project. *Voices: A World Forum for Music Therapy, 12*(3).

DeNora, T. (2000). *Music in everyday life*. Cambridge: Cambridge University Press.

Enge, K. E. (2015). Community music therapy with asylum-seeking and refugee children in Norway. *Journal of Applied Arts and Health, 6*(2), 205–215.

Fuglestad, S. (2016). Kreativ musikkverkstad – ein arena for profesjonell utvikling i barnevernspedgagoutdanninga? In K. Stensæth, V. Krüger, & S. Fuglestad (Eds.), *I transitt – mellom til og fra. Om musikk og deltagelse i barnevern. Series from the Centre for music and health* (Vol. 9, pp. 81–106). Oslo: Norwegian Academy of Music.

Hallstead, J. (2013). "It just makes you feel really good": A narrative and reflection on the affordances of musical fandom across a life course. In L. O. Bonde, E. Ruud, M. Skånland, & G. Trondalen (Eds.), *Musical life stories. Narratives on health Musicking. Series from the Centre for music and health* (Vol. 6, pp. 75–94). Oslo: Norwegian Academy of Music.

Helsedirektoratet. (2013). *Nasjonal faglig retningslinje for utredning, behandling og oppfølging av personer med psykoselidelser*. Oslo: Helsedirektoratet.

Helle-Valle, A. (2016). Samfunnsmusikkterapi i barnehagen - en anledning til utvidet omsorg? In K. Stensæth, V. Krüger, & S. Fuglestad (Eds.), *I transitt - mellom til og fra. Om musikk og deltagelse i barnevern. Series from the Centre for music and health* (Vol. 9, pp. 143–115). Oslo: Norwegian Academy of Music.

Henderson, G. (2015). Public health approaches to social isolation and loneliness: A health and wellbeing directorate seminar director, wellbeing and mental health: https://www.bristol.gov.uk/documents/20182/34732/Public-Health-Approaches-to-Social-Isolation-and-Loneliness-Part-1_0.pdf/cb17ee30-0810-4587-bca6-1b589c597479

Hense, C., & McFerran, K. S. (2017). Promoting young people's musical identities to facilitate recovery from mental illness. *Journal of Youth Studies*. https://doi.org/10.1080/13676261.2017.1287888.

Holt-Lundstad, J., Smith, T. B., Baker, M., Harris, T., & Stephenson, D. (2015). Loneliness and social isolation as risk factors for mortality: A meta-analytic review. *Perspectives on Psychological Science, 10*(2), 227–237. https://doi.org/10.1177/1745691614568352.

House, J. (2001). Social isolation kills, but how and why? Editorial comment. *Psychosomatic Medicine, 2001*(63), 273–274.

House, J.S., Landis, K.R., & Umberson, D. (1998). Social relationships and health. *Science*, *241*, 540–545.

Hunt, M. (2005). Action research and music therapy: Group music therapy with young refugees in a school community. *Voices: A World Forum for Music Therapy, 5*, 2.

Imrie, R., & Hall, P. (2001). *Inclusive design: Designing and developing accessible environments*. London: Spon Press.

Kristiansen, D. L. (2016). Å dele av seg selv. Om sangdeling som musikkterapeutisk metode i arbeid med rusmiddelavhengighet. In K. Stensæth, V. Krüger, & S. Fuglestad (Eds.), *I transitt – mellom til og fra. Om musikk og deltagelse i barnevern. Series from the Centre for music and health* (Vol. 9, pp. 211–230). Oslo: Norwegian Academy of Music.

Krüger, V. (2016). Musikk som ressurs for ungdommers livslange læringsbehov – et tverrfaglig eksempel fra skolen og barnevernets praksis. In K. Stensæth, V. Krüger, & S. Fuglestad (Eds.), *I transitt – mellom til og fra. Om musikk og deltagelse i barnevern, Series from the Centre for music and health* (Vol. 9, pp. 63–80). Norwegian Academy of Music: Oslo.

Krüger, V., & Stige, B. (2015). Between rights and realities – Music as a structuring resource in the context of child welfare aftercare. A qualitative study. *Nordic Journal of Music Therapy, 24*(2), 99–122. https://doi.org/10.1080/08098131.2014.890242.

Krüger, V., & Strandbu, A. (2015). *Ungdom, Musikk, Deltagelse – Musikk i forebyggende arbeid*. Oslo: Universitetsforlaget.

Krüger, V., Strandbu, A., & Stige, B. (2014). Musikkterapi som ettervernstiltak i barnevernet, deltagelse og jevnalderfellesskap. *Norges Barnevern, 2*(3), 78–93.

Krüger, V., Bolstad, N., & Stige, B. (2016). Hvorfor og hvordan utvikle musikkterapi i barnevernet? *Norges barnevern, 3–4*(93), 322–338. Oslo: Universitetsforlaget. https://doi.org/10.18261/issn.18911838-2016-03-04-14.

Laiho, S. (2004). The psychological functions of music in adolescence. *Nordic Journal of Music Therapy, 13*(1), 47–63.
Lau, J. (2016). *Social intelligence and the next generation*. Report rom National Citizen Service. London: King's College London and NCS.
Law, M. (2002). Distinguished scholar lecture: Participation in the occupations of everyday life. *American Journal of Occupational Therapy, 56*(6), 640–649.
MacDonald, R. A. R., Kreutz, G. & Mitchell, L. A. (Eds.). (2012). *Music, health and wellbeing*. Oxford: Oxford University Press.
MacDonald, R. (2013). Music, health, and well-being: A review. *International Journal of Qualitative Studies on Health and Well-Being, 8*(10) (no paging). https://doi.org/10.3402/qhw.v8i0.20635.
Malloch, S., & Trevarthen, C. (Eds.). (2009). *Communicative musicality: Exploring the basis of human companionship*. Oxford: Oxford University Press.
Mandal, A.B., & Bergset, L.-J. (2016). Musikkterapi for å skape samanheng i fosterbarn sine nettverk. In K. Stensæth, V. Krüger, & S. Fuglestad (Eds.), *I transitt – mellom til og fra. Om musikk og deltagelse i barnevern. Series from the Centre for music and health* (Vol. 9, pp. 143–156). Oslo: Norwegian Academy of Music.
McFerran, K. S. (2010). *Adolescents, music and music therapy: Methods and techniques for clinicians, educators and students*. London: Jessica Kingsley Publishers.
McFerran, K. S. (2016). Contextualising the relationship between music, emotions and the well-being of young people: A critical interpretive synthesis. *Musicae Scientiae, 20*(1), 103–121. https://doi.org/10.1177/1029864915626968.
Nordenfelt, L. (1991). *Hälsa och värde*. Stockholm: Bokförlaget Thales.
Roaldsnes, M. (2016). Musikk som avleiing og kjelde til positive emosjonar. Om deltaking i musikkgruppe for einslege mindreårige flyktningar. In K. Stensæth, V. Krüger, & S. Fuglestad (Eds.), *I transitt – mellom til og fra. Om musikk og deltagelse i barnevern. Series from the Centre for music and health* (Vol. 9, pp. 157–172). Oslo: Norwegian Academy of Music.
Ruud, E. (1997). *Musikk og identitet*. Oslo: Universitetsforlaget.
Ruud, E. (2010). *Music therapy: A perspective from the humanities*. Gilsum: Barcelona Publishers.
Ruud, E. (2016). *Musikkvitenskap*. Oslo: Universitetsforlaget.
Saarikallio, S. (2016). Music as emotion regulation. *Musicae Scientiae, 20*(1), 10. https://doi.org/10.1177/1029864916629856.
Saarikallo, S., Baltazar, M., & Västfjäll, D. (2017). Adolescents' musical relaxation: Understanding related affective processing. *Nordic Journal of Music Therapy, 26*(4). https://doi.org/10.1080/08098131.2016.1276097.
Sbarra, D. A. (2015). Introduction to the special section. *Perspectives on Psychological Science, 10*(2).
Skånland, M. (2013). *A technology of well-being: A qualitative study on the use of mp3-players as a medium for musical self-care*. PhD-thesis. Oslo: Norwegian Academy of Music.
Skarpeid, G. (2009). Daglig musikklytting. In E. Ruud (Ed.), *Musikk i psykisk helsearbeid med barn og unge, Series from the Centre for Music and Health* (Vol. 2, pp. 131–151). Oslo: Norwegian Academy of Music.
Small, C. (1998). *Musicking: The meanings of performing and listening*. Hanover: Wesleyan University Press.
Stene, I. (2009). Musikkterapi som gruppetilbud for ungdom. In E. Ruud (Ed.), *Musikk i psykisk helsearbeid med barn og unge, Series from the Centre for music and health* (Vol. 2, pp. 85–100). Oslo: Norwegian Academy of Music.
Stensæth, K., & Jenssen, D. (2016). "Deltagelse" – en diskusjon av begrepet. In K. Stensæth, V. Krüger, & S. Fuglestad (Eds.), *I transitt – mellom til og fra. Om musikk og deltagelse i barnevern* (Vol. 9, pp. 15–36). *Series from the Centre for music and health*. Oslo: Norwegian Academy of Music.
Stensæth, K. (2017). *Responsiveness in music therapy improvisation. A perspective inspired by Mikhail Bakhtin*. Dallas: Barcelona Publishers.
Stensæth, K., & Næss, T. (2013). "Together!" RagnaRock, the band and their musical life story. In L. O. Bonde, E. Ruud, M. Skånland, & G. Trondalen (Eds.), *Musical life stories. Narratives*

on health Musicking. Series from the Centre for music and health (Vol. 6, pp. 263–288). Oslo: Norwegian Academy of Music.

Stensæth, K., Holone, H., & Herstad, J. (2014). PARTICIPATION: A combined perspective on the concept from the fields of informatics and music and health. In K. Stensæth (Ed.), *Music, health, technology and design, Series from the Centre for Music and Health* (Vol. 8, pp. 157–186). Oslo: Norwegian Academy of Music.

Stensæth, K., Krüger, V., & Fuglestad, S. (2016). *I transitt – mellom til og fra. Om musikk og deltagelse i barnevern. Series from the Centre for music and health* (Vol. 9). Oslo: Norwegian Academy of Music.

Stige, B., & Aarø, L. E. (2012). *Invitation to community music therapy*. New York: Routledge.

Storø, J. (2016). Møteplasser for deltagelse, egenutvikling... og musikk, i ettervern. In K. Stensæth, V. Krüger, & S. Fuglestad (Eds.), *I transitt – mellom til og fra. Om musikk og deltagelse i barnevern, Series from the Centre for music and health* (Vol. 9, pp. 49–62). Oslo: Norwegian Academy of Music.

Strandbu, A. (2011). *Barnets deltakelse, hverdagslige og vanskelige beslutninger*. Oslo: Universitetsforlaget.

Strandbu, A., Krüger, V., & Lorentzen, M. (2016). Musikkteater som barneverntiltak. Identitet, fritid og kvalifisering til videre deltagelse. In K. Stensæth, V. Krüger, & S. Fuglestad (Eds.). *I transitt – mellom til og fra. Om musikk og deltagelse i barnevern. Series from the Centre for music and health* (Vol. 9, pp. 231–250). Oslo: Norwegian Academy of Music.

Trondalen, G. (2016a). Musikkterapi som anerkjennelse. En mor-barn gruppe innenfor rammen av barnevernet. In K. Stensæth, V. Krüger, & S. Fuglestad (Eds.), *I transitt – mellom til og fra. Om musikk og deltagelse i barnevern. Series from the Centre for music and health* (Vol. 9, pp. 107–126). Oslo: Norwegian Academy of Music.

Trondalen, G. (2016b). *Relational music therapy: An intersubjective perspective*. Dallas: Barcelona Publishers.

Tuastad, L. (2016). "Til alle som falt...en gang." Erfaringar frå rusforebyggjande konsertføredrag for konfirmantungdomar. In K. Stensæth, V. Krüger, & S. Fuglestad (Eds.), *I transitt – mellom til og fra. Om musikk og deltagelse i barnevern. Series from the Centre for music and health* (Vol. 9, pp. 191–210). Oslo: Norwegian Academy of Music.

Turkle, S. (2011). *Alone together: Why we expect more from technology and less from each other*. USA: Basic Books.

Uchino, B. N., Cacioppo, J. T., & Kiecolt-Glaser, J. K. (1996). The relationship between social support and physiological processes: A review with emphasis on underlying mechanisms and implications for health. *Psychological Bulletin, 119*, 488–531.

van der Kolk, B. A. (2017). Developmental trauma disorder: Towards a rational diagnosis for children with complex trauma histories. *Psychiatric Annals, 35*(5), 401–408.

WHO. (2001). *International classification of functioning, disability, and health children and youth (ICF-CY)*. Geneva: World Health Organization.

Part III
Music as a Prophylactic Resource: Examples of Projects and Initiatives

Chapter 10
Bonding Through Music: Music Therapy as Health Promotion for Mothers and Children at a Public Health Clinic

Tora Söderström Gaden and Gro Trondalen

Introduction

The birth of a child is normally associated with happiness and joyful circumstances. The baby represents hope for the future, and the parents are motivated to do their very best to provide their child with optimal conditions for development. Having a child has even been said to offer a psychological rebirth to the parents (Nugent 2014), though it is also a period of great vulnerability that is associated with an increased risk of depression and other mental health problems that can impact parents' capacity to care for their child (Slinning et al. 2010). This impact encompasses the parents' availability, presence, and sensitivity toward the child, all of which are crucial to a healthy development of the emotional bond between parent and child (Hansen 2012). A good start for both parents and their children thus has a big impact on the child's development and lifelong health:

> Some have compared a child's evolving health status in the early years to the launching of a rocket, as small disruptions that occur shortly after take-off can have very large effects on its ultimate trajectory. Thus, "getting things right" and establishing strong biological systems in early childhood can help to avoid costly and less effective attempts to "fix" problems as they emerge later in life. (Harvard University 2010, p. 5)

The first 3 years of a child's life are significant for a wide range of biological, psychological, and social functioning. The child's environment, both literal and figurative, at this time can impact lifelong emotional health, stress regulation, immune system effectiveness, and health-related behaviors (Harvard University 2010, p. 5).

T. S. Gaden (✉)
Akershus University Hospital, Oslo, Norway
e-mail: Tora.Soderstrom.Gaden@ahus.no

G. Trondalen
Center for Research in Music and Health, the Norwegian Academy of Music, Oslo, Norway
e-mail: Gro.Trondalen@nmh.no

L. O. Bonde, T. Theorell (eds.), *Music and Public Health*,
https://doi.org/10.1007/978-3-319-76240-1_10

Supporting new parents and their children should therefore be an important aspect of public health work worldwide. The Norwegian Institute of Public Health (2011) ranked measures to help parents as sixth out of the ten most important areas within health promotion and prevention, noting that efforts to support and empower parents at this vulnerable time will have significant positive outcomes for public health even beyond the immediately involved families themselves. Nobel Prize winner and Professor of Economics James J. Heckman (2011) has demonstrated through his work that investing in childhood development at the earliest age—and preferably at birth—has the greatest value for future generations. Music therapy programs in neonatal intensive care have demonstrated how music therapy can meet complex needs of both parent and child at the very beginning of life, for instance, by reducing maternal anxiety and stabilizing infant vital signs (Bieleninik et al. 2016; Loewy et al. 2013; Loewy 2015).

The present text, however, is based on a research study carried out at a public health clinic (PHC) within the Norwegian primary child services.[1] Nine first-time mothers and their infants attended a weekly music therapy program for 2 months (Gaden 2015). The group of participants was non-clinical, meaning that neither mothers nor children were known to belong to any particular at-risk population or to have specific health challenges. "Sing & Grow" is an Australian music therapy initiative offered to families at risk. The primarily governmental funded program offers children aged from birth to five and their parents an "opportunity for shared music participation and reflective discussion that encouraged the celebration of daily parenting achievement, big and small" (Teggelove 2017, p. 153). Notably, services by this program include home visits and family sessions, in addition to group workshops, and programs lasting 8 weeks. "Sing & Grow" aims at offering a non-stigmatized music environment. A similar music therapy initiative, also Australian, is "The Music Together Program" available to all families with preschool children (0–4 years) within a community, not just those identified as "at risk" (MacKenzie and Hamlett 2005). The present study shares the focus of working with a non-clinical group but differs in other ways such as how the group was targeted at very young babies during their first year of life. In addition, the present initiative was performed at a public health clinic, and was free of charge, instead of in a congregational setting where participants payed to attend. Hence, the present initiative was offered within what could be called a "neutral and official" context, which might be of significance to the parents.

As mentioned, however, the present study relied on the underlying assumption that becoming a mother for the first time is a major life event that exposes psychological vulnerability and brings with it an increased need for support. This is in line with the infant researcher Daniel Stern's (1995) description of the new mother's complex psychological reorganization of the self as she enters the "motherhood constellation"—a new mind-set and self-system that depend on a host of new

[1] The research study was approved by the Norwegian Social Science Data Services (NSD). The sessions were conducted in accordance with the code of ethics for Norwegian music therapists. The participants gave their informed consent.

cultural, personal, and contextual factors. The mother's sense of self primarily organizes around her baby and their mutual sharing of being together. The mother occupies with feeding and protecting her child while exploring and developing her maternal behavior and emotional bonding. It is also interesting to notice that recent literature points to the fact that a decrease in well-being surrounding the circumstances of the first birth influences parents' choice of having another child or not. Margolis and Myrskylä (2015) therefore argue that policy-makers worried about low fertility should be aware of factors that affect the well-being of new parents.

The goal of this mother-infant research project was to develop a low-threshold music therapy program to be carried out at the public health clinic in line with its official mandate, which, in addition to monitoring children's development, is to support pregnant women and parents of infants and toddlers and to observe and strengthen the interaction between parents and their children (Helsedirektoratet 2014). This new program was intended to support and facilitate positive parent-child interaction through songs and musical activities, similar to the examples of programs mentioned above. These are all informed by the hypothesis that participation in program would contribute to and promote good health for both mothers and children. Sharing musical experiences would contribute to the development of a positive relationship between mother and child. A music group consisting exclusively of first-time mothers was chosen in the interests of providing them with a social arena and network that would together establishing a safe space for sharing experiences and knowledge, including their worries and insecurities about motherhood, as well as the positive parenting experiences that were emerging in their everyday lives.[2]

This article consolidates knowledge regarding how music therapy may promote families' and parent-infant health and thereby how music therapy can contribute to public health in a broader sense. The questions it addresses are as follows: (1) How can music therapy contribute to supporting new parents and their children during the first year of life? (2) How can a carefully designed music therapy program for first-time mothers and their children promote health?

Theoretical Perspectives

Health Musicking

Over the last 20 years, authorities in the areas of both health and culture have become increasingly interested in using cultural resources for health promotion. Art forms and therapies including music and music therapy have, in fact, already been activated as health-promoting activities (Ansdell and DeNora 2016; MacDonald et al. 2012; Ruud 1998). While music has many aspects, it is in this case seen

[2] We invited only mothers to participate due to our previous experience from a pilot, where only mothers signed up to participate. A possible explanation would be that mothers in Norway as a rule take the first period of the parental leave, and the project included only very young babies.

primarily as a means of communication. In addition, it boasts therapeutic potential in that it attunes to human regulatory systems—both internal neurochemical, hormonal, and metabolic processes and external engagements with people and objects in the world (Trevarthen and Malloch 2000).

Sometimes, mothers and children express themselves through music; other times, they listen to music together. Either way, music facilitates the contact, and the musical relationship is at the forefront of the joint creation, within a dyad or a group (Bjørkvold 1989; Trolldalen 1997). This musical process of relational bonding involves mutual affective exchange, including imitation, and cross-modal exchange[3] of form and contour via face-to-face contact within a defined timeframe (Trevarthen 1980). Through songs and other musical activities, mothers can express their feelings about and connection to their children (Jacobsen and Thompson 2016a, 2012): "Music, with dance and all the expressive arts, offers a direct way of engaging the human need to be sympathized with—to have what's going on inside appreciated intuitively by another who may give aid and encouragement" (Trevarthen and Malloch 2000, p. 11). Songs such as lullabies connect us to ourselves, including our bodies, minds, emotions, and memories (Bonnár 2014). Jacobsen and Thompson, for example, drew attention to emerging characteristics in working with families (Jacobsen and Thompson 2016b). As theory was concerned, resource-oriented and family-centered belief systems, with the resilience and flexibility of the family, were at the forefront. Core terms like affect attunement and attachment seem to be vital, as was communicative musicality. In the music therapy approaches, the music therapist focused on the quality of interactions between family members, as she respectfully focused on the family's knowledge, wishes, and resources, in addition to promoting the sustainable use of music. The musical relationship in these settings is interactive in nature and emerges from improvisational activities; it is multilayered and unfolds in the here and now, though, paradoxically, it is not bounded by these qualities: "The phenomenal musical relationship then emerges as an art form—a field of relational lived experiences—emerging from an inborn, communicative musicality" (Trondalen 2016, p. 89).

Music therapy in Norway is rooted in a humanistic tradition (Ruud 2010) and values context and the power of the emergent musical relationship. User participation and a focus on personal resources are cornerstones of a music therapeutic approach (Trondalen et al. 2010) situated within the broader field of discourse known as "music and health," which encompasses all of the ways people use music to promote health (see Chap. 8 by Stige and Chap. 9 by Stensæth). Bonde (2011) proposes a model for "health musicking" that views music as a social phenomenon and activity, in line with musicologist Small's (1998) term "musicking." Health musicking links music to concepts such as the formation of identity, body awareness, personal experience of meaning, interaction with other people, self-expression, and

[3] Cross-modal perception involves interaction between two or more sensory modalities. For example, the mother says "aaahhh" to her baby, and the baby's response is stretching her legs. There is an obvious correspondence in expressive exchange, however, in different modalities. It is a contact on the basis of form and contour within a time span (Hansen 2010).

meaning making. Bonde (2011) equates it with affirmative and corrective emotional and relational experiences that can be used to regulate one's state of mind or promote well-being.

People, then, may sing, participate in a choir, dance to music, compose songs, play precomposed music, or join a band as part of a reflexive strategy to improve their health and well-being. Ruud even suggests that "musicking" may serve as a "cultural immunogen" (Ruud 2013a). It is therefore relevant to elaborate upon the unique health potential that can be found in musical experiences as tools for developing one's agency and empowerment and as a resource or form of social capital (Bourdieu 1998) in building one's social network and locating meaning and coherence in life. Health musicking involves the whole human being and has both social and individual aspects (Bonde 2011). Bonde's model illustrates both the complexity and the richness that underpin the many ways that people use music to promote health in their daily lives. Using music to "perform" health positions it as a productive or generative concept as well as an "added-value" experience.

Health Promotion and Prevention

In 1948, the World Health Organization (WHO) presented their infamous definition of health as "a state of complete physical, mental and social well-being and not merely the absence of disease or infirmity" (Huber et al. 2011, p. 235). This groundbreaking gesture overrode the negative definition of health as the "absence of disease" and also expanded it to encompass both mental and social health. Ultimately, that definition faced criticism for making health an almost unachievable state; nevertheless, it remains unchanged to this day.

Health experts have since introduced nuance to the definition based on certain dynamic factors such as resiliency, mastery, and the ability to manage, maintain, and restore one's well-being, aligning health with "the ability to adapt and to self-manage" (Huber et al. 2011, p. 343). Within this understanding, health promotion is all about contributing to the experience of positive health. The priorities are to feel good, to master daily life, to be able to cope with challenges, and to experience well-being (Huber et al. 2011). Health is also now a relationship rather than a condition and one that involves both physical and mental aspects, as well as existential well-being. Physical and mental health are not discrete entities but instead continuous, so that bodily functions influence one's mental state and vice versa (Sigurdson 2008). Likewise, health is now understood in relation to one's experience rather than the objective measurement of one's bodily functions. All of this context implies that the music therapist will downplay the pathological aspects of one's life in favor of its healthy dimensions, in order to allow for self-actualization and the search for meaning within a relational context (Trondalen 2011). Ideas about health are also culturally informed and vary widely across societies: "If the role of cultural systems of value in health is ignored, biological wellness can be focused on as the sole measure of wellbeing, and the potential of culture to become a key component in health maintenance and promotion can be eroded" (Napier et al. 2014, p. 1607). WHO (2016), in addition, emphasizes that health promotion is a process that enables people to increase control over and access means of improving their health, moving this work beyond a focus on individual behavior to encompass a wider range of both social and environmental interventions and factors.

The terms "health promotion" and "prevention" are often used somewhat interchangeably, but they are different. According to the Bureau of Norwegian Health and Social Affairs (2004), health promotion includes efforts to improve the conditions for good health and promote prosperity, well-being, and mastery of daily challenges and stresses. Prevention, however, includes efforts aimed at reducing illness or death and their associated risk factors. Improving one's health-related conditions and practices, then, may also mitigate the possibility of ill health. For this reason, we propose that the practice of music therapy groups within the context of a primary health service center qualifies as a clear example of public health work focused on health promotion.

Method

Design

The research study was performed with a non-clinical group of nine first-time mothers between 27 and 36 years old, and their infants between 12 and 18 weeks old, who attended weekly music therapy group sessions at their local public health clinic (PHC). The PHC is a municipal service administered by Norwegian primary child services that offer families free health-care services emphasizing health promotion and prevention. The service has a very high attendance rate across the population. Its three divisions address children from 0 to 5 years old and their families, primary schools, and adolescents, respectively (Helsestasjonstjenesten 2017). This project was realized in collaboration with the first division.

The research design was qualitative. After completing a program of eight sessions over a 2-month period, the mothers participated in individual semi-structured interviews (Kvale and Brinkmann 2009). The individual interviews were carried out either at the PHC or in the participants' homes, with a semi-structured interview guide as an outline. Accordingly, the interviews also included themes that arose from the group, both through the participants' shared reflections and observations made by the music therapist. The participants were invited to share their expectations prior to participation, why they had signed up, and their anticipated outcome. Furthermore, they were asked to describe their children's experiences of being a part of the group, the musical experiences, as well as their personal experiences. Examples would be what they enjoyed the most and whether anything had been difficult or challenging. They were also encouraged to reflect upon if, and if so how, what it meant to them that the group consisted solely of first-time mothers. In addition, they were asked if gathering at the public health clinic had any special significance to them. The questions also included the mothers' use of music with their child in their daily lives, both before, during, and after participation in the program. Finally, the mothers were asked if, eventually how, they shared their experiences with their partners and/or family and friends.

The interviews were recorded and transcribed, before the transcripts were analyzed in accordance to interpretative phenomenological analysis (IPA) (Smith et al. 2009). IPA is an approach to qualitative research rooted in phenomenology and hermeneutics, with an ideographic focus where each participant's individual experiences situated in a given time and context are explored. IPA presents with a stepwise procedure where interview transcripts are repeatedly read and interpreted and then coded in detail into various themes to grasp the richness of the participants' experiences. Eventually themes are grouped in broader categories which still aim to capture the variety of patterns of experience. In IPA, research is viewed as a collaboration between participant and researcher, as both the participants themselves reflect upon and interpret their own experiences during the interview, before the researcher continues the interpretation process. Hence, the analysis can be seen as a collaborative product, reflecting both the participant's original statements, reflections and interpretation, as well as the researcher's interpretation and contextualization of these (Smith et al. 2009).

The Music Therapy Sessions

The music therapy program was carefully designed to meet the needs of the mothers and infants (Robb et al. 2010; Schwartz 2013; Edwards 2011). The sessions lasted approximately 60–90 min. Prior to the music activities, the participants were invited to arrive early to talk and take care of their babies' immediate needs, such as feeding and changing diapers, so that they would be in the best condition possible for participating in the music. The content of the sessions consisted of improvised musical activities with an emphasis on song, touch, and movement. A music therapist participated in all the activities, mainly by modeling them using a teddy bear rather than offering verbal directions. This allowed the mothers to focus on the interaction with their children and the music rather than on the instructions. During the sessions, the music therapist and the mothers also reflected upon the experience in the here and now. The mothers were encouraged to share their own early childhood memories of music, along with their uses of music in their daily routines with their children. This was to increase awareness of how music could impact both mothers and children while also highlighting significance of early musical memories to add meaning to the act of musicking together with their child. Further on, research demonstrates that practices of care are passed on to subsequent generations (Slade et al. 2005), and hence it is probable that being sung to as a child increases chances of using song and music as part of parenthood and practice of care as parents later on. Early childhood memories of music are often related to bodily experiences and feelings of intimacy and safety, and they very often involve the presence of family members or caregivers (Bonnár 2014; Ruud 2013b).

Results and Reflections

In the semi-structured interviews, analysis showed that the mothers described several ways in which participation in the music therapy program had a positive impact on a personal level, on their relationships with their children, and on their everyday lives. In the following results are presented and followed by some reflections. The headings refer to the four subordinate themes resulting from data analysis, namely, personal outcomes, relationship with the child, social outcomes, and transferability to everyday life.

Personal Outcomes

The mothers' personal outcomes relate both to experiences of well-being during the musical interaction in the group and to personal growth through an extended repertoire, both musically and in their new role as mothers.

They described that their musical experiences affected them immediately in the here and now and made them feel connected to their inner selves and their bodies. One observed how the music experiences brought out a strong feeling of presence:

> I forgot everything else. We were supposed to switch off our phones and all that … So yes, I was completely present. I felt really good. Almost like a meditative state of mind. Yes, it was just a time to take a break from it all. Everything is just put aside, right? All the disturbing elements.

Several of the mothers experienced the weekly music therapy sessions as a valuable contrast to their hectic everyday lives and to "disturbing elements" such as cellphones, technology, and social media. Shared musical experiences offer points of intersection that connect the participants to one another and to different parts of themselves. The program focused on song and the use of voice, an instrument that has the unique quality of being part of the body. No other musical expression is as effective as singing in regulating the body at a physiological level. When one sings, one breathes more deeply, one's heart rate slows, and one's central nervous system is calmed (Austin 2008).

The voice is deeply personal, as well, as it connects to our identity and the audible realization of who we are. Sounds and especially music effectively connect to our bodily emotions and memories (Trondalen 2016). In the present program, the musical repertoire mainly consisted of well-known children songs. Singing these songs offered the mothers a way to revisit their own pasts:

> Everything comes right back to you when you have your first child. You keep thinking: "Wow, yes, I remember *that* song."

As they reflected upon and explored their own early musical memories, the mothers discovered a sense of purpose and meaningfulness that motivated them to keep singing to their children:

> It makes you realize how important the songs from your childhood have been ... with all the memories they bring back now. And then you think of how you yourself really should sing for them [your children] too, having all these good memories about it yourself.

For some of the mothers, it had been a long time since they had sung, and they expressed a lack of confidence or dislike of their own voices. It felt good, then, to find that one does not need a perfect singing voice to have a meaningful musical moment with a baby:

> It's really nice to see her respond, because I don't have a particularly nice singing voice. I just sing with the voice I've been given, but still it's nice to experience that my baby doesn't care about how my voice sounds. She is just fascinated by me doing something else or making sounds.

> When *I* sing, she goes completely calm ... It made me feel more like a mother. The kind of mother that you are used to from your childhood – one who sings for you.

Their children's responses to their voices thus gave these women renewed confidence in their abilities as both mothers and singers.

Relationship with the Child

These outcomes were related to how shared musical experiences between mother and child led to them discovering new sides of their child, as well as providing mother and child with a different way of being together that facilitated communication and bonding.

Several of the mothers described how their children's unique personalities and competences revealed themselves during the musical experiences. Seeing their children actively participating in both musical and social interactions was indeed exciting to the mothers, who enjoyed watching these babies grow, develop, thrive, and enjoy music throughout the program:

> I see that she uses her voice and has rhythm and movement in her body. I see that she's here, *participating*—and that's been extremely nice. I've discovered things about her that I might not have seen if we hadn't participated in the program.

Infants are born with a sensitivity and an inner motivation to engage and interact with other human beings. Each infant is unique in its personality, temperament, and inner rhythm (Stern 1985). Musical experiences can also provide new means of communicating for both mother and baby, which is welcome:

> When they are that young, it can be so hard to reach them, to find a way of communicating together. But when it comes to music, it feels like it provides you with something different. Music is easy, playful, and positive.

Mothers noted that they had been told that it was good to talk to a baby. At the same time, some of them found it hard to continue talking and trying to connect when they were alone with their baby and felt they were not getting any response.

As opposed to this, song and music provided them with another means of communication:

> Sometimes it is hard. When you don't get any response from them … That's why songs and riddles are so helpful … It's such a nice way to communicate with her, and through singing I also get more used to hearing my own voice.

Through shared musical experiences, both mother and child developed a mutual language, experienced and mastered adequate interplay, and created joint memories together. Participating in music therapy brought them closer together and strengthened their attachment and general feelings of belonging:

> All of a sudden, you get eye contact, she starts smiling if you're playing around a little bit… In the end, I really feel that it has strengthened our relationship a lot.

Mothers found that positive experiences shared with their children fortified the bond between them while also enhancing the feeling of belonging to one another. They were astonished that their babies preferred their voices, which also contributed to a feeling that they were special to their own children.

Social Outcomes

Social outcomes concern how having a social activity to look forward to during the week—where they usually spent much time alone with their baby—was experienced as particularly meaningful by the mothers. Also, the group provided an arena where they could share experiences, exchange advice, and provide support for each other.

As mentioned, the group consisted exclusively of first-time mothers, whose shared life situation seemed to promote a sense of a safe space and a feeling of community. The mothers found that the group gave them a social and supportive network and a feeling of belonging, which was very important, as isolation, loneliness, and the lack of a support network can pose a big threat to one's sense of well-being and quality of life. It was valuable to the mothers to meet others who were struggling with the same things, in order to share experiences and offer advice and support:

> When you are a first-time mother, everything is completely new. I feel so vulnerable. One is afraid—and everything should be done in certain ways … there is so little pressure here in the group, which affects the whole atmosphere. It is nice to know I do something good for my child.

Transferability to Everyday Life

Lastly, the content of the sessions—the musical activities and the songs—was transferable to their everyday lives and became part of daily routines with their children. The mothers described using songs to soothe and calm, as well as to distract and help out in challenging situations:

> I sing for her in situations where she gets upset or unhappy. For example, while I'm dressing her I sing, and through this I'm able to kind of distract her. Instead of her getting fussy or starting to cry, she's attentive to my singing … I know how to make her happy.

Clearly, the music therapy group supported personal self-agency (Stern 2000), which empowered families, when the mothers related and transferred their experiences to their spouses and partners at home.

> After I started singing more for her I've also heard her daddy sing some of the same songs. He gets inspired, because he has seen her giving so much response when I sing, so I guess it's fun for him to try it out himself too.

Song and music became a way of being together for the whole family, as well as a helpful tool for both parents in challenging situations.

Discussion

Parenthood and practices of care are constantly changing as a result of new knowledge, family constellations, social and cultural conditions, and values. Nevertheless, one principle seems to be a constant: one's appreciation of oneself and one's child as uniquely present to one another at a personal level. The music therapy program supported and strengthened an emergent motherhood for these first-time mothers. Such support of a sustainable "motherhood constellation" is linked to healthy and creative personal development (Stern 1995).

Today's parents have unlimited access to information via the internet, where parenting blogs, apps, and various fora supply them with advice that is only a click away. On one hand, we believe that this can enhance parental confidence and independence. On the other hand, we argue that this can encourage the passive consumption of other people's competence, experience, and advice while one's own innate capacities and resources may remain unrealized.

In all cultures, at all times, crying babies have prompted both rhythmic rocking movements and the use of the voice in the adult (Bjørkvold 1985; Malloch and Trevarthen 2009). Many parents in today's Western societies, on the other hand, tend to replace their own singing with recorded music and, in general, sing less for their children than ever before (Creighton 2011). The mothers in this study reported that they limited their singing due to lack of experience and/or shyness, and this inclination would impact whether and how parents integrate health musicking into their daily parenting practice. In the music therapy group, of course, the mothers introduced preverbal communication and musical elements into their interaction with their children without necessarily being aware of how important this humming, singing, rocking, and general playfulness were to the babies (Bonnár 2014; Dissanayake 2008). The mothers also reported sharing these experiences with their spouses and partners, hence, transferring their personal experiences into these families' daily lives.

Shared musical experiences provide a framework within which both mothers' and infants' communicative musicality can be put into play. Song, movement, and touch are musical forms of expression that align with an infant's mode of communication (Malloch and Trevarthen 2009; Stern 2000). When interaction and communication take place on the infant's terms, parents are better able to recognize and applaud the infant's social competence and initiative. In our opinion, parents should be provided with a nuanced picture of what music can offer their child, within which music's relational qualities—its unique way of connecting us to ourselves and others while also promoting immediate contact and unique moments of communication—should be acknowledged and emphasized.

The music therapy program did not seek to achieve specific goals but rather to explore the process of musicking itself. Its shared musical experiences empowered the relationship between mother and infant, because the music provided a framework that fostered intimacy, communication, and interplay, in turn offering meaning and support to an emergent motherhood. In each group session, the mothers and babies made music together by attuning themselves to music and their participation in it. Parental singing represents a unique form of communication that conveys emotional content in a way that is completely unlike recorded music. The attendant joy and laughter were visible and audible to each mother-baby pair and to the other participants. The group as such, that is, offered both support and fun, but also meaning, because meaning constitutes itself in one's phenomenal musical experiences within a culture, including the present music therapy program. In short, playing is for real (Trondalen 2016, p. 120).

This research project was carried out in a Norwegian cultural context, within which there resides a rich tradition and repertoire of both folk music and children's songs and nursery rhymes (Ruud 2015). Prompting parents to use song and music in the care of their children might serve to counteract what we see as a negative trend where many parents tend to sing less to their children or even replace their own singing with recorded music with the result that valuable practices of health musicking as part of the child's care and expression of parental love might get lost. The PHC is an important arena in this work that could evolve from a place where health is biologically "measured" and "controlled" to a place where parents find support in establishing a foundation for their child's health, development, and quality of life using music as an important health-promoting factor (Ruud 2015). Low-threshold music programs led by music therapists at local public health clinics represent a useful form of introducing parents and their children to an enjoyable, accessible, and low-cost means of promoting their health.

Limitations, Implications, and Ethics

The present study is based on a qualitative research methodology, with a small and homogenous sample of participants, within the context of a Norwegian public health clinic. Consequently, the result might not be directly transferable toward the population at a general level. Nevertheless, the study highlights the public health clinic as a suitable context for music therapy work with parents and infants, and the participants' experiences provide in depth knowledge on what kind of health benefits participation in such a group may provide.

The results of the present music therapy study at the public health clinic has shown potential to support mother-infant interaction and bonding, as well as empowering first-time mothers in their new role. Accordingly, such a supporting health initiative should be further investigated, not least in a larger scale. In future research, it would be useful to look at a similar music therapy group using a quantitative or mixed-method research design, such as a randomized controlled study or a multi-site study, including a bigger number of public health clinics participating. It would also be useful to develop music therapy guidelines for interventions, in addition to a broad range of research methodology, while also actively involving participating parents in the research process.

There are also ethical considerations to be made when working with parents and their newborn children. In this case, there should be a particular focus and awareness toward the formation of intuitive parenting and development of the intimate caregiving relationship. The music therapist should be especially aware to attune herself toward the parents in a way that support and empower a good enough parenthood. It might, however, be argued that supporting and encouraging the discovery of the child and the building on their relationship through the nonverbal and active interaction of musicking might be less intrusive than other forms of parent support and guidance. Supporting parenthood through the power of musicking is an investment for life.

Conclusion

This chapter presents perspectives on how music therapy groups at local public health clinics can serve to promote parents' and infants' health practices and consequently contribute to public health in a broader sense. Research has shown that efforts to support and empower new parents have significant positive impacts on public health both within and beyond the families in question (Harvard University 2010; Heckman 2011). In addition, both health and cultural authorities have become increasingly interested in using cultural resources for health promotion. With the music therapy program presented here, music was a means of communication between mothers and children, as well as a regulator of both the mother's and the infant's physiological, psychosocial, and emotional states of being.

The mothers in this study found that the carefully selected program at the PHC enabled them to uncover new aspects of their baby's individual personality, competences, and preferences. When parents discovered themselves able to use songs and music to soothe, calm, and distract these children in otherwise fraught situations, they became more confident, both musically and with regard to motherhood and parenting. Their experiences in the group were then transmitted to spouses and partners, hence, becoming part of their daily life activities.

The group of first-time mothers soon built a social, supportive network that contributed to the sense of belonging and acceptance. On basis of their feedback, we conclude that participating in the music therapy program contributes to promoting the health of both mother and child. The mothers said that music-making allowed for good feelings and an experience of well-being while providing them with tools to cope with the challenges and stresses of daily life (Huber et al. 2011). We believe that offering music therapy programs to new parents and their children at public health clinics is an investment in Norwegian public health for generations to come.

References

Ansdell, G., & DeNora, T. (2016). *Musical pathways in recovery: Community music therapy and mental wellbeing*. Farnham: Ashgate.

Austin, D. (2008). *The theory and practice of vocal psychotherapy: Songs of the self*. London: Jessica Kingsley Publishers.

Bieleninik, Ł., Ghetti, C., & Gold, C. (2016). Music therapy for preterm infants and their parents: A meta-analysis. *Pediatrics, 138*, e20160971.

Bjørkvold, J. R. (1985). *Den spontane barnesangen—vårt nye morsmål*. Oslo: J. W. Cappelens Forlag AS.

Bjørkvold, J. R. (1989). *Det musiske menneske*. Oslo: Freidig Forlag.

Bonde, L. O. (2011). Health musicking—music therapy or music and health? A model, empirical examples and personal reflections. *Music and Arts in Action, 3*(2), 120–140.

Bonnár, L. (2014). *Life and lullabies: Exploring the basis of parents' lullaby singing*. PhD dissertation. Oslo: Norges musikkhøgskole.

Bourdieu, P. (1998). *Practical reason. On the theory of action*. Cambridge: Stanford University Press.

Creighton, A. (2011). Mother-infant musical interaction and emotional communication: A literature review. *Australian Journal of Music Therapy, 22*, 37–56.

Dissanayake, E. (2008). If music is the food of love, what about survival and reproductive success? *Musicae Scientiae, 12*, 169–195.

Edwards, J. (2011). *Music therapy and parent-infant bonding*. Oxford: Oxford University Press.

Gaden, T. S. (2015). *Samspill, sang og trygge bånd. En gruppe førstegangsmødres opplevelser av et musikkterapitilbud på helsestasjonen sammen med barna sine*. MA thesis. Oslo: Norges musikkhøgskole.

Hansen, B. R. (2010). Affektive dialoger. Fra regulering til mentalisering. In V. Moe, K. Slinning, & M. Bergum (Eds.), *Håndbok i sped-og småbarns psykiske helse* (pp. 116–136). Oslo: Gyldendal Norsk Forlag.

Hansen, B. R. (2012). *I dialog med barnet. Intersubjektivitet iutvikling og psykoterapi*. Oslo: Gyldendal Norsk Forlag AS.

Harvard University, Center on The Developing Child. (2010). *The foundations of lifelong health are built in early childhood*. Retrieved from http://www.developingchild.harvard.edu

Heckman, J. J. (2011). The economics of inequality: The value of early childhood education. *American Educator, 35*(1), 31.

Helsedirektoratet. (2014). *Samfunnsutvikling for god folkehelse*. Retrieved from www.helsedirektoratet.no

Helsestasjonstjenesten. (2017). *Helsestasjonens oppgaver*. Retrieved from http://www.helsestasjonstjenesten.no/helsestasjonens-oppgaver/

Huber, M., Knottnerus, J. A., Green, L., Van Der Horst, H., Jadad, A. R., Kromhout, D., et al. (2011). How should we define health? *BMJ: British Medical Journal, 343* (Jul 26 2), 235–237:d4163.

Jacobsen, S. L., & Thompson, G. (Eds.). (2016a). *Music therapy with families: Therapeutic approaches and theoretical perspectives*. London: Jessica Kingsley Publishers.

Jacobsen, S. L., & Thompson, G. (2016b). Working with families. Emerging characteristics. In S. L. Jacobsen & G. Thompson (Eds.), *Music therapy with families. Therapeutic approaches and theoretical perspectives* (pp. 309–326). London: Jessica Kingsley Publishers.

Kvale, S., & Brinkmann, S. (2009). *Interviews: Learning the craft of qualitative research interviewing*. Los Angeles: Sage.

Loewy, J. (2015). NICU music therapy: Song of kin as critical lullaby in research and practice. *Annals of the New York Academy of Sciences, 1337*(1), 178–185.

Loewy, J. V., Stewart, K., Dassler, A. M., Telsey, A., & Homel, P. (2013). The effects of music therapy on vital signs, feeding, and sleep in premature infants. *Pediatrics, 131*(5), 902–918. https://doi.org/10.1542/peds.2012-1367.

Macdonald, R., Kreutz, G., & Mitchell, L. (Eds.). (2012). *Music, health, wellbeing*. Oxford: Oxford University Press.

MacKenzie, J., & Hamlett, K. (2005). The music together program: Addressing the needs of 'well' families with young children. *Australian Journal of Music Therapy, 16*, 43–59.

Malloch, S., & Trevarthen, C. (2009). *Communicative musicality: Exploring the basis of human companionship*. Oxford: Oxford University Press.

Margolis, R., & Myrskylä, M. (2015). Parental well-being surrounding first birth as a determinant of further parity progression. *Demography, 52*, 1147–1166. https://doi.org/10.1007/s13524-015-0413-2.

Napier, A. D., Ancarno, C., Butler, B., Calabrese, J., Chater, A., Chatterjee, H., et al. (2014). Culture and health. *The Lancet, 384*, 1607–1639.

Norwegian Health and Social Affairs. (2004). *Kommunenes helsefremmende og forebyggende arbeid i helsestasjons- og skolehelsetjenesten.* Veileder til forskrift av 3 April 2003 nr. 450.

Norwegian Institute of Public Health. (2011). *"Bedre føre var ..." Psykisk helse: Helsefremmende og forebyggende tiltak og anbefalinger* (Report 2011:1). Retrieved from http://www.fhi.no/dokumenter/1b2e13863a.pdf

Nugent, J. K. (2014). The newborn period: Where hope and happiness meet. In J. Gomes-Pedro (Ed.), *Working with infants and families: Towards a science of happiness*. Canterbury: Gulbenkian Press.

Robb, S. L., Carpenter, J. S., & Burns, D. S. (2010). Reporting guidelines for music-based interventions. *Journal of Health Psychology, 16*(2), 342–252.

Ruud, E. (1998). *Music therapy: Improvisation, communication and culture*. Gilsum: Barcelona Publishers.

Ruud, E. (2010). *Music therapy. A perspective from the humanities*. Gilsum: Barcelona Publishers.

Ruud, E. (2013a). Can music serve as a "Cultural Immunogen"? An explorative study. *International Journal of Qualitative Studies of Health and Well-Being, 8*, 20597. Retrieved from https://www.ncbi.nlm.nih.gov/pubmed/23930988.

Ruud, E. (2013b). *Musikk og identitet*. Oslo: Universitetsforlaget.

Ruud, E. (2015). Musikkterapi—Fra antikkens tenkning til moderne helseprofesjon. In E. Ruud (Ed.), *Fra musikkterapi til musikk og helse: Artikler 1973–2014. Bind 1* (pp. 107–133). Oslo: NMH-Publikasjoner.

Schwartz, E. K. (2013). Early intervention. In M. R. Hintz (Ed.), *Guidelines for music therapy practice in developmental health (chapter 2)*. Gilsum: Barcelona Publishers.

Sigurdson, O. (2008). Vil du bli frisk? In G. Bjursell & L. V. Westerhäll (Eds.), *Kulturen och hälsan. Essäer om sambandet mellan kulturens yttringar och hälsans tillstånd* (pp. 189–218). Stockholm: Santérus Förlag.

Slade, A., Grienenberger, J., Bernbach, E., Levy, D., & Locker, A. (2005). Maternal reflective functioning, attachment, and the transmission gap: A preliminary study. *Attachment and Human Development, 7*(3), 283–298.

Slinning, K., Hansen, M., Moe, V., & Smith, E. (2010). *Håndbok i sped-og småbarns psykiske helse*. Oslo: Gyldendal akademisk.

Small, C. (1998). Musicking. The Meanings of Performing and Listening. Hanover: Wesleyan University Press.
Smith, J. A., Flowers, P. & Larkin, M. (2009). *Interpretative Phenomenological Analysis: Theory, Method and Research*. London: Sage Publications
Stern, D. N. (1985). *The interpersonal world of the infant: A view from psychoanalysis and developmental psychology*. New York: Karnac Books.
Stern, D. N. (1995). *The motherhood constellation*. New York: Basic Books.
Stern, D. N. (2000). *The interpersonal world of the infant: A view from psychoanalysis and developmental psychology*. New York: Basic Books.
Teggelove, K. (2017). Building stronger families through music. Sing&Grow Groups Programs for families at risk. In S. L. Jacobsen & G. Thompson (Eds.), *Music therapy with families. Therapeutic approaches and theoretical perspective* (pp. 152–172). London: Jessica Kingsley Publishers.
Trevarthen, C. (1980). The foundations of intersubjectivity: Development of interpersonal and cooperative understanding in infants. In D. R. Olson (Ed.), *The social foundations of language and thought* (pp. 316–342). New York: Norton.
Trevarthen, C., & Malloch, S. N. (2000). The dance of wellbeing: Defining the musical therapeutic effect. *Nordic Journal of Music Therapy, 9*(2), 3–17.
Trolldalen, G. (1997). Music therapy and interplay: A music therapy project with mothers and children elucidated through the concept of "Appreciative Recognition". *Nordic Journal of Music Therapy, 6*(1), 14–27.
Trondalen, G. (2011). Music is about feelings: Music therapy with a young man suffering from anorexia nervosa. In T. Meadows (Ed.), *Developments in music therapy practice: Case study perspectives* (pp. 434–453). Gilsum: Barcelona Publishers.
Trondalen, G. (2016). *Relational music therapy: An intersubjective perspective*. Dallas: Barcelona Publishers.
Trondalen, G., & Stensæth, K. (Eds.). (2012). *Barn, musikk og helse* (Vol. 2013:1). Oslo: Norges musikkhøgskole.
Trondalen, G., Rolvsjord, R., Stige, B. (2010). Music therapy in Norway: Approaching a new decade. In *Voices: A world forum for music therapy*. Retrieved from http://www.voices.no/?q=country-of-the-month/2010-music-therapy-norway-approaching-new-decade
World Health Organization. (2016). *What is Health Promotion?* Retrieved from http://www.who.int/features/qa/health-promotion/en/

Chapter 11
Singing for a Better Life: Choral Singing and Public Health

Anne Haugland Balsnes

Introduction

Some years ago I attended a public health conference focusing on "Promoting health in everyday settings".[1] The programme included presentations about lifestyle themes such as obesity, nutrition, exercise, drugs and so on. The most important arenas for health-promoting work seemed to be the workplace, schools, nurseries and other local institutions. Leisure activities were, on the other hand, almost completely absent, apart from one session about health-promoting work at sports clubs. Beyond my own presentation about choral singing and another about "gardening" in prisons (which can possibly be considered as a cultural activity in terms of an expanded concept of culture), the arts and cultural expressions were absent at the conference.

Despite appearances at the aforementioned conference, there has nonetheless been a certain reorientation within the health services in recent years where the focus has changed from a concentration on one-sided curative activities and moved towards preventive and health-promoting ones (Fugelli 1998), which target lifestyle and health habits in general. WHO has, for several decades, recognised that health factors exist to a great extent outside the health sector (Stige 2006), and the Ottawa Charter from 1988 establishes that "health promotion is not just the responsibility of the health sector, but goes beyond healthy lifestyle to well-being".[2] Health problems in the Nordic countries are today to a lesser extent concerned with threats from poverty and deficiency diseases. For many people, poor life quality is connected to non-material health threats such as a lack of fellowship, faith, meaning

[1] Downloaded 27.11.2013 from http://www.hive.no/about-vuc/conferences/nhprc/.

[2] Downloaded 27.4.2013 from http://www.who.int/healthpromotion/conferences/previous/ottawa/en/.

A. H. Balsnes (✉)
Ansgar University College, Kristiansand, Norway
e-mail: balsnes@ansgarskolen.no

L. O. Bonde, T. Theorell (eds.), *Music and Public Health*,
https://doi.org/10.1007/978-3-319-76240-1_11

and hope (Fugelli 1998). These are factors which belong more to other contexts than medical ones.

In this article, I will discuss how a leisure activity such as choral singing, can promote public health.[3] Choral singing is, in many countries, the one musical activity which attracts the most people.[4] In Norway, for example, more than 5% of the adult population sing in choirs (Balsnes 2009). As a result, there should be good reasons to study the choral movement as a health institution. There is a growing body of research on choral singing from a health perspective (see for instance surveys of existing research in Clift et al. 2008; Clift et al. 2010; Balsnes 2017). The results tend to point towards choral singing as a health-promoting resource.

The main objective of the article is to discuss how choral singing can function as a resource in public health work. The question will be discussed on the basis of individual choir members' accounts about their choral singing drawn from various qualitative studies where participant observation and qualitative research interviews were used as collection techniques. The selected "cases" are very different; the singers in question have varied backgrounds and current life situations, and the choirs they sing with are distinct with regard to purpose, level, repertoire and methods. Nevertheless, all these singers explain that participation in choral singing has contributed to a better life and thus improved their health. Their stories are analysed on the basis of questions such as: What is the meaning of the singing for the participants? What factors associated with choral singing might allow for a beneficial impact connected to health? What conditions must be present? And what characterises the benefits that choral singing provides?

Theoretical perspectives from music therapy and music sociology will illuminate the discussion, in particular Ruud's categories suggesting the relationship between music and quality of life and Stige's concept of *health musicking*. Through studying accounts of choral singing in light of relevant theoretical perspectives, we might begin to approach an understanding of the phenomenon choral singing and public health, with its focus on health promotion and improving of lives on the population level.

---

[3]The article is an extended and adapted version of "Korsang for et bedre liv: Kor, identitet og helse" [Choral singing for a better life: Choir, identity and health], which is published in the popular science publication, *Hjertesproget. 16 forsknings- og praksisbaserede studier af sangens egenskaber, vilkår og virkning* (2016, Herning: Videncenter for sang, p. 165–180).

[4]American reports also document that in the USA, choral singing scores highest in terms of the total number of participants in arts and cultural activities, see Chorus America (2009). *How children, adults, and communities benefit from choruses: The chorus impact study.* Downloaded 15.4.2013 from https://www.chorusamerica.org/advocacy-research/chorus-impact-study.

Musicking and Health

What is the connection between musical performance in general and health? The term *musicking* characterised by Small (1998) is helpful. Musicking points towards music as practice and process – as something we *do* instead of as an *object* (ibid.). It is thus not choral singing as an object that we are studying here, but singing as an activity or process. Music can do little "on its own". It is therefore not the music in itself which has an effect on people but rather the *paramusical aspects,*[5] as music therapist Gary Ansdell terms it (2013) – everything that is "round" the music. The power of the music becomes effective when it is united with the people who practice it and the context in which it is practiced. Music has no inherent qualities by its nature – it is dependent on person and situation (Ruud 2013). Music is thus not a means, but works as a *catalyst* (Batt-Rawden 2010). Music offers something (affordance), which must be used by people (appropriation) (DeNora 2013). If music is used in a health setting, we can then talk about *health musicking* (Stige 2012, see Chap. 8 by Stige). The concept of health musicking is developed within the music therapy field but is nevertheless relevant in other areas. One central thought is that health effects are not given but develop in the situation at hand: "What music can afford in relation to health [...] grows out of the relationships established in each case" (Stige 2012, p. 184). Health musicking focuses on mobilising musical and paramusical resources in the service of health. There is also a focus on everyday life, rather than clinical settings, since it is in everyday settings that the majority of health factors exist:

> [H]ealth musicking can be understood as the common core of any use of music experiences to regulate emotional or relational states or to promote well-being, be it therapeutic or not, professionally assisted or self-made. (Bonde 2011:140)

Thus, health musicking can take place in a therapy situation as well as at choir rehearsal. Stige suggests the following definition of the term:

> Health musicking could be defined as the appraisal and appropriation of the health affordances of the arena, agenda, agents, activities, and artefacts of a music practice (Stige 2012: 186)

– in this case, choral singing. These different categories are closely related to one another and affect each other (cf. Andsell's notion, the "ecology" of health musicking, 2013). The arena in which the musical activity takes place affects both the agenda and the activity and vice versa. All of these elements represent both possibilities and limitations. Below, I will discuss the possibilities connected to the various choral practices accounted for here.

When we are to discuss the connection between music and health, we must also examine which health term is most relevant. Biomedicine – which defines health as the absence of illness – is too narrow. Health is not a static condition, but a time-based

[5] Stige also prefers "paramusical" to the alternative, "extramusical", to avoid the misleading impression that things are either "totally" musical or "not musical" (2012, 186).

dimension which is dependent on material, social and symbolic aspects. Both health and illness are thus complex and diverse phenomena. How different states arise, are identified, perceived and treated, as well as their social consequences, are all dependent on cultural context (DeNora 2013, p. 137). From a humanist perspective, one can thus see health as an *experience* – an experience of well-being and meaning in life. Health becomes more of a resource and not the goal of existence itself – but rather a means to achieve, for example, well-being. Health is therefore something, which is created through processes and developed through relations. With such a constructivist or interpretative health term, music professor Even Ruud believes that we can equate health with life quality (Ruud 2010). Other terms expressing the same thing include *contentment*, *well-being* or simply a *good life*. In this way, health is understood as having a positive emotional life, life coping skills, good social contacts and relations to others, as well as a feeling of meaning and life coherence in one's existence (ibid.). Ruud claims that *musicking* (Small 1998) can help people with such needs and thus work to promote health. All initiatives and activities that can contribute to these areas will thus advance public health.

From such a perspective, we see that identity and health are closely related. A person can be healthy in a medical sense but nonetheless feel that they do not have a good life. Another person might suffer from one or more illnesses but nonetheless have a perception of herself which accommodates *more* than just the illness. Identity is constructed through the narratives that we tell about ourselves (Ruud 2013). These narratives might well have connections to musical events and experiences (ibid.), for example, accounts of choral singing, on which is the focus here. Music is not just an effective tool in the work to create and mark out identity, but it can also function as a powerful and practical means to repair, confirm and maintain identity (ibid.). Musical practice can therefore work *transformatively* – when we make music, we become what we do (DeNora 2013, p. 141). I become a choral singer when I sing in a choir – and become part of a social and musical fellowship when I sing together with others. Music-making can also have a *maintenance* function which contributes to coherence in life when upheavals such as illness, war or other circumstances strike. In the following section, I will present examples of what choral singing has meant for people in different life situations.

Choir Accounts, Methods and Materials

The accounts referred to here are brought in from five different studies carried out in connection with the research project, *Choral Singing for a Better Life*. Four of the studies are case studies where participant observation and interviews were used as the main data collection techniques. In addition, an interview study was carried out. All the interviews in the different studies were semi-structured. They were recorded and transcribed. Ethical guidelines were followed in the sense that all informants gave their consent and were anonymised. The analysis is inspired by the method of thematic coding (Kvale 2007).

In all the studies, the main purpose was to investigate the meaning of choral singing. The aim with including several studies in the material for this article is to obtain a broad range of material for analysis, which includes a variety in the types of choir and participant. From the total collected interview material, seven singers who each belong to a different choir were selected and will have their voices heard here. The seven singers were chosen because they were either typical examples or "extreme cases" within the respective studies. Their stories function as "embedded cases" or "mini-cases" within the main cases (Stake 2000). The primary criterion in the selection process was simply the opportunity *to learn* (ibid.).

The different studies and the chosen interviewees with their respective choirs are presented below.[6]

Diana in a Community Choir (Case Study 1)

Between 2003 and 2006, 17 amateur choral singers from *Belcanto,* a "typical" Norwegian mixed choir with a strong local connection, were interviewed (Balsnes 2009, 2012). Everyone who enjoys singing is welcome to join the choir, which has a varied repertoire. Most of the performances take place within the local community, in church or in other local contexts. At the time the choir was studied, it had existed for around 20 years, and many of the members had been involved from the start. The 17 interviewed singers, who made up around a third of the total membership, were selected in order to achieve the greatest possible degree of variation with respect to age, gender, place of residence and time of experience with the choir.

In the same period, participant observation of choir practices, performances, tours and festivals was carried out, as well as document analysis of Board minutes, annual reports and reviews.

The story about Diana is derived from this study. Diana struggled with poor self-confidence as a result of bullying during her childhood. As an adult, she moved to a new place and met some people who sang in the choir, *Belcanto*. Diana was persuaded to join them and has since been "hooked" on choral singing.

Louis in a Multicultural Gospel Choir (Case Study 2)

KIA Multicultural Gospel Choir (The KIA Choir) in Kristiansand was studied during spring 2012 (Balsnes and Schuff 2013; Balsnes 2017). The choir is part of KIA – which stands for "Kristent interkulturelt arbeid", or "Christian Intercultural Work". The organisation works for "more multicultural fellowship, and for equality, consideration and friendship between all people in Norway, regardless of cultural

[6] There is no space for a very detailed overview of methodological aspects within the framework of this text. I refer, therefore, to other texts which further clarify methods and material.

background, language or religion".[7] In addition to the choir, KIA arranges homework help, communal dinners, cultural cafes, courses and seminars, football training, women's groups, parties, free Norwegian teaching, trips, camps and more. The choir is thus just one of several activities that KIA is involved in. The choir was made up of a mix of foreign students, au pairs, refugees, asylum seekers, foreign workers and majority Norwegians "who think that this is a good place to be". The leaders were committed to creating a warm and inclusive atmosphere and extra follow-up for those who needed it – whether it be, for example, a phone call on the day of rehearsal, transport or something else. The social aspects of the rehearsals were given extra emphasis. The choir practiced an open door policy – everybody was welcome regardless of musical expertise, ethnicity, social background or status. Norwegian was the primary spoken language at the rehearsals. The leaders tried hard to speak "simply" in order to help newcomers to understand the information given. The choir sang with three vocal parts (soprano, alto and tenor) – never four or more. Sheet music was not in use – instead, audio files were posted online. The repertoire consisted of three categories of song: Original compositions based on stories about immigrant experiences, songs from the participants' home countries and songs that are known worldwide, such as *Amazing Grace*. Many of the songs had simple choreography.

The study focused on members with backgrounds as refugees – a group which is often vulnerable with regards to both mental and physical health (Dalgard et al. 2006). Two of the choir's directors and five singers from different countries were interviewed (Peru, Chile, Rwanda, Burundi, Germany). Four of these came to Norway as refugees – one of them is Louis, whose story is told here. The singers were contacted after suggestions from the leaders, and the selection criterion was simply "somebody who could have a story to tell". Participant observation of rehearsals and concerts was also conducted for one semester.

Louis survived the genocide in Rwanda but lost his whole family. He fled and came to Norway. When he arrived, he came into contact with the organisation KIA and joined the KIA choir.

Jonas in a Singing at Work Project (Case Study 3)

In the spring of 2013, a workplace choir was set up as five lunch hour sessions for an organisational unit with around 30 employees (Balsnes and Jansson 2015). The department had just been created as a merger of two smaller departments, and the manager positioned the project as an opportunity to come together as a team in a different and enjoyable way. The choir sessions took place over a period of 2 months in the lunch lounge. A few participants were experienced singers, but most had no choral singing experience. Three pieces of a cappella music were prepared. After the last session, the choir performed for two senior managers from headquarters,

[7] Downloaded. 21.6.2011 from www.kianorge.no/om-oss. (Translated by the author.)

and a separate meeting was organised for everyone to collectively reflect on the experience. The meeting was run as a large-scale group interview, with a combination of questions and round-table answers and a free-flowing exchange of views. The duration of the meeting was a little more than an hour. The sound recording of this meeting serves as the primary data. The group interview was complemented by a questionnaire about the individual experiences of the project, as well as notes from participant observation.

Jonas was one of the employees who took part in the workplace singing. He had never sung in a choir before and had certainly never thought he ever would. When he heard that there was to be a choir project at work, he was sceptical. He was afraid that it would steal time away from more important tasks. In addition, he was somewhat shy about his own voice.

Peter in a Senior's Choir (Case Study 4)

Sølvstrupene [*The Silver Voices*] is a choir for senior citizens organised by the theatre and concert house, *Kilden* [*The Source*], in Kristiansand. The lower age limit is 62 years. There is no upper age limit, but the average is around 70. Social aspects are important. The doors open half an hour before rehearsals for coffee. Many stay afterwards to eat lunch and attend Kilden's concert programme series for senior citizens, which takes place the same day. The choir performs together with the local symphony orchestra on May 17th (the Norwegian Constitution Day) and at Christmas concerts, in addition to other concerts and tours. The choir sings classical works with orchestra, as well as a "lighter" repertoire. The repertoire is meant to provide challenges, as well as being suitable for people without choral experience. The instructors work hard to make the choir sound as good as possible. They explain that they maintain demands on the singers, which are just as high as they would do in a choir with younger members.

During the autumn of 2013, six singers in the choir were interviewed (Balsnes 2017). An informative e-mail about the study was sent to the choir members, in which those who wished to be interviewed were encouraged to make contact. Five people volunteered, and a sixth person was asked to supplement the selection in order to achieve more breadth in the material. In addition, informal conversations with organisers, instructors, the members' council and other singers were held during the observation period, which lasted one semester. Both rehearsals and performances were observed. Peter represents the senior choir in this article. He joined the choir after becoming a pensioner.

Cristin in a Chamber Choir, Lisa in a Professional Choir and Lily in a Church Choir (The Interview Study)

In addition to the four case studies, an interview study was carried out with choral singers who had various chronic illnesses (Balsnes 2014). Six singers with different diagnoses (schizophrenia, post-traumatic stress syndrome, rheumatism, leukaemia and complex syndromes) were interviewed in the course of 2012. They were recruited in different ways; two of them were already part of a different research project and were asked to contribute in this study, two were singing in a choir the author used to conduct some years ago, one was contacted after advice from a conductor college and one after having read his story in a newspaper. All of them sang in choirs at a high level – one of them in a professional choir. The aim was to find out what choral singing can mean for people with particular health challenges, such as chronic illnesses. Cristin, Lisa and Lily, whose stories are told here, were amongst the participants in this interview study.

Cristin grew up in a musical family, but it was not before she moved to a larger town as an adult that she started singing in a choir – which she enjoyed hugely. Some years later, she became sick and was diagnosed with chronic leukaemia. The illness affects her life significantly, but she has nonetheless managed to continue with choral singing. At the moment, she is part of an ambitious chamber choir with an audition, which often sings together with the city's symphony orchestra.

Lisa too grew up in a musical family, singing, playing the piano and participating in choirs from her childhood onwards. Lisa suffers from post-traumatic stress disorder following incidents during her childhood and adolescence, and she has had a difficult life. She has not had the health to develop herself as a soloist, but she has managed to keep singing in choirs. She is a member of a professional choir which maintains a high standard and has a demanding repertoire.

Lily has many health problems, both physical and mental, following a difficult childhood and adolescence. She has been in and out of different health institutions throughout her life. Despite many stays in hospital, she has maintained her membership of a church choir in her local community. When she is at home, and during good periods, she sings with this choir.

The Meaning of Choral Singing

In the following analysis, the presentation of the meaning of the choral singing for the chosen interviewees is structured into four subpoints, inspired by Ruud's categories pertaining to the correlation between music and health (Ruud 2010). This is followed by a summary of each singer's story.

Choral Singing Provides Positive Experiences and Emotional Peaks

Jonas was sceptical towards the choir project he was "subjected to" at work. Afterwards, he was surprised by the results. He explains how there was much smiling, laughter and humour in connection with the rehearsals and the experience of not taking each other so seriously. He felt rested and filled with energy afterwards.

Also Diana explains with enthusiasm how she found choral singing to be:

> I get feelings like this sometimes which can't be described. They're so good and it feels like they come from the heart – yes it is hard to explain. But it gives me so much. I get raised up, yes, really lifted up. Singing is both curative and health promoting. Oxygen flows round the whole body. I get a lift. I am certain that – to a man – some of them might be just so tired when they turn up at the choir, but when they come out again after the practice, I think they almost fly home.

Cristin describes singing in a choir in this way:

> It's almost like falling in love – you get drawn to it. It's almost like a little fix when you get to choir and get a lot of good stuff and can start singing – great sounds, great works.

Cristin also gives expression to the sense of how singing in a choir is a form of recreation, which is particularly important when life is affected by illness:

> Right from the time when I became sick, choral singing has been a kind of recreation. There is a sense of sorrow in losing your health. But [at the choir] I forget almost what is unpleasant. I can distance myself from illness and pain when I sing. I live in my own musical bubble.

Some of the interviewees relate experiences with choral singing, which can almost be described as transcendent, spiritual, holy or mystical experiences, as Lisa does:

> The moments when it happens: when you are singing, it happens in the moment and it is big. I am a part of something big. It's about intensity. Fellowship. Affirmation. Solidarity. We have to play in the same team. It's almost something mystical. It's buzzing. It touches on something you aren't quite able to explain…

Choral Singing Provides Experiences of Mastery and Contributes Towards Empowerment

Experiences of mastery are important for a good life. Just as "healthy" people have arenas such as work, school and so on, it is extra important that people who, for different reasons, remain on the outside of such contexts, have alternative arenas for mastery. Cristin, who cannot have a regular job because of her illness, relates the following:

> It is important to have something to do yourself if you can't work normally – that you have something to do and can get feedback that you are of use. That is what *that* is about. That you can see it yourself, too – that "I can fix this". If not it's a total failure.

She also explains that it is important to get to show off one's musical talent – "I can show that I can stand on a stage as I did before, give something to people – there is something magical in that".

Louis explains what happened in the choir a short time after he had come to Norway: "They asked if someone had a song from their home country. I volunteered and taught the others a song. They loved it! They sang the song everywhere". For Louis, who came as a refugee to Norway without speaking a word of Norwegian, the feeling that he had something to contribute with, and could be a resource rather than just a recipient of help, was powerful. The same thing happened to Lily when she came home to her choir between hospital stays:

> It was something I could attend – you see that I can read music, and it was good to have an alto there who can just about sight-sing and who is a support. I felt that it was nice for me to be a part of this because I could contribute with something. They wanted to have me in the middle so that those on each side could hear – sometimes someone would want me at the back, but the director wanted to have me at the front. They learnt something from listening to me – they learnt more from it than from practicing at home. This was a good thing for me. I became a point of support for the alto group. I was not used to being a kind of fulcrum anywhere. It was very peculiar. [...] I felt, even if I felt awful in every way, that I had an assignment in the choir and it was fun carrying it out – very nice and very good.

The feeling of meaning something also applies to other singers. Peter explains that in *Sølvstrupene* "I feel that I have a place which I can fill. It is good to mean something again". He also expresses happiness at still being able to receive instruction in later adulthood, and the feeling of achievement, particularly when they have practiced new material for a long time. Jonas, of whom it cannot exactly be said that he belongs to a vulnerable group, admits nevertheless that he was outside his comfort zone, particularly at the beginning of the choir project. However, he became steadily more confident and ended up with a positive feeling of mastery. He was extremely proud when they performed the songs at the end of the project.

Diana explains that being able to hide away in the crowd was the factor which meant that she, as a shy person, dared to sing. Through her activity in the choir, Diana has grown as a person, she explains. After having conquered her initial fear, she was also persuaded to join the board. After a few years, she ended up becoming chair of the board. Now, Diana dares to do things she has never done before. She, who never dared raise her voice in social gatherings, never allows an opportunity to speak pass her by now.

For refugees in a new country, integration is important for a good life, and the key to integration is learning the language. Louis explains how participation in choral activities has taught him Norwegian "without being with a teacher":

> It isn't formal – it's friends. There's nobody who laughs at you if you pronounce something strangely. *The KIA Choir* is the perfect place. There is a mixture of Norwegian and Spanish and English – it is so relaxed. And when you are going to learn a new choir song, you first have to learn its meaning.

Choral Singing Provides Fellowship and Social Contacts

Choral activities create social meeting places, which are a starting point for fellowship and networking. A choir can be described as a resource and care centre (Balsnes 2009). Diana, as a newcomer in the village, quickly gained new friends after joining the local choir. She relates one episode 3 months after she started. Diana had been absent for a while in connection with a hospital stay. At the first rehearsal back after her absence, the following happened:

> There I was on my way in on crutches, and when they saw me, they were so happy to see me that they even began to clap, and then the tears came. I thought it was fantastic to be welcomed in that way – I had been there three months, barely that.

Some of the choral singers whose stories are told in this study have a particular need for a social network – they are singers with health challenges who do not have the opportunity to be employed in regular work. Cristin explains:

> It is wonderfully good to come down and meet these people. They are my friends. Most people in the choir have another job where they meet other people. But I don't. I actually have to seek out people if I want to meet anybody. It is not a given with me that I would meet people other than my family at home. The choir is my social connection point.

Becoming a pensioner is a change in life, which can mean less social fellowship. When the possibility of fellowship with work colleagues disappears, one can risk becoming isolated if one does not find new social arenas. Peter explains that starting out in a choir has served to extend his everyday routine, not just on the day of the rehearsal, but because he was invited to take part in other activities by contacts in the choir. Louis, who came to Norway alone as a refugee, also found the choir to be of vital importance:

> We all shared being in a new country – new hopes, fears… we were in the same shoes, but from different countries, studying together, studying the same language, a new way of life – that's why I felt at home among the choir members.

Louis calls the choir his new *family*. What is special with being a part of this choir is the *way* in which people are together. It is not just any club or society. Gathering for an activity which itself promotes interdependence and fellowship is important. Lily's social life has been complicated as a result of her history of illness. However, she explains how choir participation has "saved" her from isolation and loneliness:

> I have always found a choir to be a part of. In retrospect, I can see why I have done it – when I was involved it was a matter of course that I did it. But it has always been this which has saved me, to put it simply. I got to know some, got a few contacts, and we met up in a different way to how it was at school or in colloquia. It was both the music, and the fact that we were doing something together. Because I never did anything together with others. I see now, though, others had clubs and followed each other around all the time.

Jonas explains that he does not think it is easy being in contexts where there is a lot of "small talk", such as lunch at work. For this reason, he found choral singing to be liberating – a different way to be together. There was great variation regarding previous choral experiences amongst participants, but in spite of this, Jonas felt that the choir project "shook them closer together". He claims that the sensitivity one needs to have in relation to the voices of others in a choir is absolutely transferable to daily cooperation in a company. He explains, amongst other things, that it was a strange experience to stand beside his manager and sing. In the choir project, everyone is equal. In a choir, all voices are just as important – the first bass is not more important than the second bass – it is simply a vocal definition. The aim of choral singing is not competition, as in sport, but cooperation and the good of the fellowship as a whole. Just as everyone is equally important, people are also dependent on each other in a particular way. Lisa elaborates on this by referring to her own situation:

> I have not been healthy enough to develop my voice enough to be a soloist, but I *have* been healthy enough to sing in a choir. Singing in a choir has been perfect for me – it is a fellowship which carries you, whilst others are also completely dependent on me. I must not let them down – I must do what must be done, but we are also together in doing this, so the result is much larger than the sum of the individual parts.

Choral Singing Contributes to Meaning and Coherence in Life

Several singers stress the importance of having a regular activity to attend, which helps to structure their lives. For Peter, "Tuesday is such a good day", he explains, "because then we have rehearsals and drink coffee afterwards. For me, the choir is something regular to attend, and it gives me an extra 'leg to stand on' – one that I didn't know that I had". Diana claims that Thursdays are "sacred", they are "the choir's day". Regular trips abroad to visit family are arranged after consultation with the choir's programme. Cristin, who suffers from a chronic illness, says:

> It has been essential for me to sing in a choir during all these years. If I hadn't had the singing, I don't know what my mental health would be like. It has managed to keep me going mentally and it has given me something to look forward to, and something regular to attend. Otherwise I don't think I would have been doing so well mentally.

For some, dramatic events have created ruptures in their lives – events such as illness or the need to flee. Choral singing, in such a situation, means that something *known* is maintained. As Lisa puts it: "I have never been 'choirless'. No matter where I have lived, there has been a choir".

In the study of the multicultural gospel choir, a representative of the local refugee health services was interviewed. He explained how the days of many newly arrived refugees are characterised by sleep problems, headaches, uneasiness, a sense of longing, loss, uncertainty regarding residence permits, dissatisfaction with long waiting periods and anxiety about the future. These do not constitute major mental illnesses, but these people are nevertheless under great mental pressure as a result of the uncertainty and because they miss everything:

> If you come straight from Africa, you know nothing of society. The only thing you recognise at the store is Coca Cola. But what's inside the boxes – which food it is you see in front of you – you have no reference points. And then you miss all the sounds, smells. Your basic safety net is shaken because you are so foreign.

In a situation like this, the nurse stresses that a choir can contribute with something known. Immigrants from Africa in particular know about singing in choirs from home:

> It something they are confident with, know, and master. It provides fellowship. Suddenly you have a new focus, instead of just focusing on these small struggles of everyday life, of arguing with the other mothers [in shared apartments], waiting for a response from UDI, constantly living with fear. Thus, you give them new content in their everyday lives – something positive to ponder and think about – and a community.

What does it mean when Diana says: "Music and the choir, it's my life!"? Or when Cristin claims that "I cannot do without music"? Such assertions say something about a deeper significance, which touches upon *meaning*. Choral singing is felt by several of the singers to be a completely *essential* part of life, which they cannot be without. Cristin's story shows clearly how "everything else" is put aside so that she can be a part of the choir:

> Even though I don't work, I still have Tuesdays where I am not in good shape, but there is no chance I will stay at home. From my point of view, the choir means so incredibly much, mentally speaking. So, I go to choir. I will have taken extra medicine so that I can go to choir. I see it as an advantage for myself – going to choir. Many times, I have not been able to sing – I just sit there, but nonetheless I am cheerful when I come home. I'll just take extra medicine, and then I'll just get on with it and meet up. When we have projects with several rehearsal days in a row – then the choir becomes life itself. And the family just has to understand that. They know that the choir is important for me. They accommodate me. They don't complain. There have been a few hints dropped here and there… "you know you have to lie down for three days afterwards…" But then I say: "You know that the choir is really important for me." "Yes, yes, I know that." So that's just the way it is. I go to choir and I can't just skip it.

Lisa describes, in a poetic manner, what a choir can mean in a particular life situation. When she was studying, she had a difficult period but was still part of a student choir:

> The choir was the golden thread. A source that was indisputable. It wasn't a result of positive thinking, cognitive efforts or expectations. It was in itself a grounding tool, a ray of light, a nerve that helped keep me there and kept me going.

Improved Lives: A Summary

Before I discuss the results, I will sum up the significance of choral singing for the different participants: Louis got a new family through *The KIA Choir*. Life became more meaningful. He was empowered and received respect. In the long term, the choir contributed to his integration in a new country and thus to a better life. He could see himself as something other than a refugee. From previously being a refugee and asylum seeker, he could now see himself as a choral singer, a soloist and a resource – one with something to contribute.

For Diana, choral singing has meant positive experiences, social belonging and an experience of mastery that have been transferrable to other areas of life. From earlier thinking that "I am amongst the most shy in society", Diana is now an assured individual who can walk with her head held high and who dares to take on tasks and leadership responsibility. She puts it like this:

> I would say that I have grown as a person. It means everything to me. Before, I was very shy and quiet and steady, and never spoke out at any meeting. And now, I have been Chair of *Belcanto* for three years [...]. I have grown hugely as a person through taking part and singing [...] So now I feel that I can walk with my head held high.

Peter went from a position of being in danger of losing his footing because of his retirement to having a new, regular anchor point in his life and something to look forward to through *Sølvstrupene*. Choir participation has meant further development, instead of stagnation. Being a pensioner has turned out to be much better than he feared it would. He says himself: "The choir has enriched my life – life has become much brighter after I joined the choir".

For Cristin, choral singing provides positive experiences, a diversion from pain and an important sense of social belonging. Her participation in the choir allows her to maintain the activities that she had before she became sick, which means a lot for her self-understanding. As she puts it: "I didn't start in a choir because I became sick. I *am* a chorister". Even with a serious, chronic illness, she has self-insight which allows her to say: "I get sick when I don't get to choir".

Jonas felt a sense of happiness, achievement and fellowship through the singing-at-work project. The project did not only change his own self-perception in relation to what he could achieve, but the working fellowship too was changed.

Lisa has, despite ill health, managed to become a professional choral singer. Different choirs throughout her whole life have been a forceful "thread" which has contributed to keeping her going.

For Lily, choral participation means that she has avoided isolation – it has given her the possibility of social fellowship. As she can read music and is a dependable singer, she has had the opportunity to see herself as a resource – an experience which life in general has not offered her.

Discussion

Through these different examples, we can see how choral singing has contributed, in different ways, to people's lives, and has resulted in bettering them. People in different life situations have different needs. Some need social contacts or something to look forward to, others need to be integrated into a new country or have a break from pain. For Jonas, who essentially had a good life, the choir nevertheless provided a greater sense of well-being at work and a feeling of achievement which was transferable to other situations. With DeNora's terminology (2013), different choral practices offer the singers various aspects (affordance), which they have grasped themselves (appropriation). Ansdell says: "Music helps when there is a good fit between what music can provide, and what we need in a given situation" (Ansdell 2015, p. 38). For the singers in question, choral singing seems to have contributed positively in different areas. What, then, characterises the benefits that choral singing provides?

A Holistic Impact

Were we to sum up the significance of choral singing as it is described above, we would see that it touches on all of the dimensions inherent in being human. Choral singing provides, therefore, a holistic effect: physical, psychological, cognitive/mental, social and existential. Ruud's categories overlap to a great extent in relation to these dimensions, but there is no specific place for the physical dimension in his model. The paragraph concerning positive experiences, however, touches on effects, which are about both emotional and physical aspects. From the physical point of view, singing affects breathing, posture and relaxation. Peter, who suffers from Parkinson's disease, explains that one of the best things he can do to relieve tension in his body is to sing. Many feel a sense of physical well-being through singing. Diana describes it as if oxygen is streaming around inside her whole body and that she – no matter how tired she was before the rehearsal – feels as if she is "flying" her way home. Research exists studying choral singing in connection with specific illnesses, for example, COPD (Skingley et al. 2013). This research shows that singing provides concrete physical effects which improve the health of such patients.

Even though unique elements of humanity are named and delineated here, it is important to point out that it is precisely the associations between the different elements, which allow choral singing to affect people in a positive way, cf. music therapist Ansdell's ecological perspective (Ansdell 2013). If we removed, for example, the social dimension – the fellowship that arises when people sing together – the effect on the other areas in question would be direct.

Salutogenic Impact

A musical activity like choral singing does not treat illnesses directly but supports the healthy aspects of the person. Even though choral singing has shown itself to have a positive impact on, as previously mentioned, for example, COPD patients' lung capacity, it cannot cure illnesses directly. What choral singing does for singers with chronic illnesses, however, is that it helps them to keep going with activities that are connected to a healthy life. Music practice can thus be seen from a *salutogenetic* perspective (Ansdell 2013) – as something that promotes health and strengthens the healthy aspects of an individual.

To the question as to whether choral singing is a therapeutic practice, leaders of the KIA choir with which Louis sings make some interesting reflections in reply. The leaders are conscious that they are not therapists and that the choral practice is not music therapy. Nevertheless, "it is just therapeutically good to be here", they say. Therapy is "being together, singing and socialising, receiving care and support".

Health Promotion Regardless of Life Situation or Type of Choir?

The choral singers who have been interviewed in the studies here are all in different life situations. The choirs they are a part of are very different with respect to arenas, agendas, agents, activities and artefacts (cf. Stige's *health musicking* concept, 2012; see also Chap. 8). An attempt at a model that explains the main differences between the choirs might look like this (Fig. 11.1):

To the left, we find choirs that were established for specific purpose, such as *The KIA Choir*, which was started with integration in mind, and the singing-at-work choir Jonas was a part of, which aimed for teambuilding in the workplace. Another type of choir in this category, which is in the process of spreading throughout Norway, is the "Sing yourself healthier" choir, which involves facilitated choral opportunities for people with mental health challenges (Bjøru 2015). When there are specific aims in mind, there is a need for particular, tailored arrangements. With participants from "the whole world", mental health issues, or those more or less "forced" to participate, repertoire and methodologies (or in Stige's word, activities and artefacts) must be adapted.

One place to the right (Fig. 11.1), in the category "open choir", are choirs with a low threshold and no audition – where everyone is welcome – as is the case with the senior choir *Sølvstrupene* (as long as one meets the age criteria), the community choir *Belcanto* or Lily's church choir. Further to the right are more ambitious choirs, perhaps with auditions, and over to the far right are professional choirs, such as the choir in which Lisa sings, where one has to almost "squeeze through the eye of the needle", so to speak, to get in. In a choir like this, the aim will simply be to perform music at a high level. We see that in this model, there is an increasing level of difficulty concerning the repertoire, and, at the same time, the choirs become more exclusive the further to the right one moves.

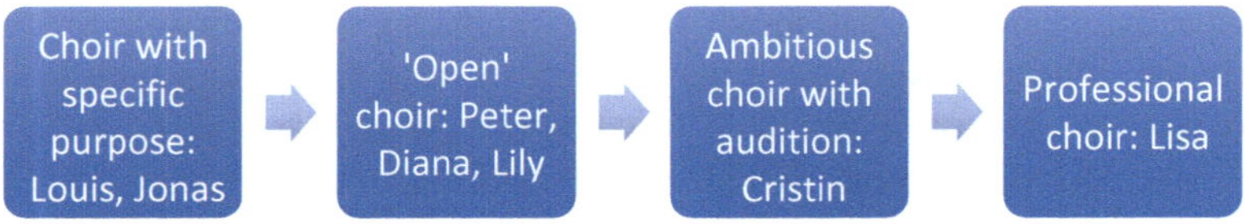

Fig. 11.1 Model for different choir types

Even though the social dimension of a choir will always be important, regardless of the level, one can assume that the emphasis on musical aspects will increase in the choirs to the right, whilst the social dimension is all the more important the further one moves to the left. We have seen, for example, that the social environment of *The KIA Choir* is particularly emphasised.

Another way to distinguish between the different choir types can be that the further to the left one moves, the more concerned one is with the *process* of choral singing, whilst to the right, it is the musical result, *the product*, which is more or equally important. Learning the music will also occur in different ways, possibly with the use of more sheet music the further one moves to the right. The level of the repertoire in the different choirs is very different, from simple three-part gospel choir songs to advanced choral music of a high level in the chamber and professional choirs. The degree of musical difficulty does not affect the participants' experiences in itself, but the music must be *adapted* to the participants. If Jonas, who had no previous choral experience, had started with the chamber choir, he would likely have had a negative feeling of mastery, and therefore the positive effects would have been absent. The same would have happened if Cristin had sung with the multicultural gospel choir. With her previous choral experience, the repertoire of this choir would have bored her and thus not provided the positive experiences she has with the chamber choir.

The significance of the fact that the repertoire and methodology must be adapted to the participants' requirements cannot be understated. There are enough people around who have had negative choral experiences. Choral singing is no quick fix and provides no automatic positive effects. Research shows that it is the repertoire, the choir leader and environment of the choir that have the most effect (Kreuz and Brünger 2012). An environment of exclusion does not provide social benefits. A conductor who does not work with the singer's well-being and mastery in mind will not create any sense of achievement for the participants. Too demanding or overly simple repertoires do not provide positive results either. Adjustment and adaptation are therefore keywords. Methodologies and repertoire (cf Stige's activities and artefacts) must be adapted to the participants in the different choirs so that the positive effects of choral singing are made possible. The activities must be adjusted to the aim of the individual choir.

Despite their obvious differences, all the choirs referred to here have nonetheless contributed to change for the participants. In some of the choirs, there are particular adjustments made for health effects, whilst in others it is the participants themselves who appropriate the health opportunities inherent in choral practice. Cristin, for example, uses choral singing as a form of self-therapy. She knows that getting to the practice means a great deal for her well-being and self-perception and therefore prioritises it to a great extent.

In light of Stige's concept of health musicking, and the elements arena, agenda, agents, activities and artefacts, we can say the following: In the selected accounts of choral singing, the arenas extend from workplaces to integration work and from churches to concert halls. The choirs have different agendas, such as integration, team-building, singing "just for pleasure" or performing music at a high level.

The participants (agents) in question are different – here we find a refugee, people with chronic illnesses, a pensioner and also people without particularly unusual characteristics. Even the learning is arranged in different ways – some use sheet music and others learn songs by ear or use sound files (artefacts). Not least, the repertoire extends from the simplest to the most advanced in choral music. Stige's health musicking covers all of this – it is about any use of music to promote well-being, whether it is with a therapeutic aim or not and whether it is professionally assisted or self-made. Choral singing can thus be said to be a good example of health musicking.

In light of the accounts, we can see that the two most important aspects to the connection between choirs and health, or better lives, if you will, are the transformative and sustaining significance of choral singing. The singing contributes to people changing their self-perception or maintaining their identity despite life affecting challenges or changes in living conditions. On this basis, choral singing can be helpful in all forms of public health work, in health promotion, in prevention work and in therapy and rehabilitation.

Limitations and Strengths of the Study

One of the strength of this investigation is that several studies are included in the material. Consequently, a broad range of material including a variety in the types of choirs and participants is analysed. A total of 36 singers were interviewed individually, and a large-scale group interview included around 30 singers (study 3). Compared with criteria for statistic survey, however, the total number is still small. With such a selection, statistic generalisability is not possible. Instead, though, it is possible to imagine a kind of analytical generalisability, which means that certain traits in the cases could be applicable in other situations (Yin 1989). Thus, the studies should provide insights and experiences that can be transferred to other contexts.

The interviewees were recruited in several ways, by advice, personal knowledge or self-recruitment after announcing. One might protest that these ways of recruiting research participants can lead to only the most eager and positive coming forward whilst removing critical perspectives. In order to avoid this, triangulation of methods has been used in the studies (except for the interview study) to obtain different perspectives. Participant observation (study 1–4), informal conversations with even more singers (1–4), survey (study 3) and document analysis (study 1 and 4) have given the opportunity to gain other perspectives. In case study 1, former members of the choir were interviewed. In the mapping of existing research in the area, also literature on negative aspects of choral singing was included (see Kreuz and Brünger 2012). Thus, the study has brought forward reliable and valid knowledge of choral singing and public health.

Looking Beyond

On the basis of the knowledge concerning choir and health, the choir movement can be said to be an important, health-promoting institution. I would like to propose the following statement: Choral singing is a cost-effective activity, which provides health effects absent of side effects. Choral singing cannot, of course, replace the National Health Service's traditional approaches, but it could function as a supplement. As music sociologist Tia DeNora puts it, music is "easy to add to an environment" (DeNora 2013, p. 138). Singing is the very easiest way to make music since the instrument that is used – the singing voice – is a part of the body. In order to make a choir, one needs a group of people, a place to be and a competent leader. The choral movement already exists as a widespread movement, and the possibility to make use of it in public health is thus great. However, as we have seen, there is a need for many different types of choir in order that people with different competencies and in different life situations should find where they thrive best. There is, not least, a need for specially adapted choir practices. One of the newest additions in Norway are the aforementioned *Sing yourself healthier* – choirs which are in the process of establishing themselves in several municipalities and which are particularly adapted for people with mental health challenges. More people can, through choirs like this, have experiences that again can give them better lives – whether we call it contentment, quality of life, well-being or simply better health.

Choirs are organised differently. Some are connected to culture and leisure departments of municipals or activity centres and others to health service or health promotion departments. Then there are all the community choirs, church choirs, chamber choirs, senior choirs, symphonic choirs, etc. To realise the potential of the choral singing movement for public health, coordination, collaboration and cooperation between the music and cultural life, voluntary organisations, public health officers, health centres, rehabilitation institutions and community workers are needed. One idea is to establish a joint network to bring forward the knowledge of choral singing and public health in local communities. This could be an important part of a salutogenic approach to public health in the communities.

Last of all, I would like to add that choral singing does, of course, have its own *intrinsic value,* even if one ignores all the "useful" things choirs can contribute to in terms of public health and other positive aims. Choral singing is, first and foremost, about *the joy of singing* and *fellowship*, and these *are* benefits in themselves.

References

Ansdell, G. (2013). Foreword: To music's health. In E. Bonde, E. Ruud, M. S. Skånland, & G. Trondalen (Eds.), *Musical life stories: Narratives on health musicking*. Oslo: NMH-publikasjoner. 2013:5.

Ansdell, G. (2015). *How music helps in music therapy and everyday life*. Surrey: Ashgate.

Balsnes, A. H. (2009). *Å lære i kor. Belcanto som praksisfellesskap*. Oslo: Unipub. (NMH-publikasjoner 2009:7).

Balsnes, A. H. (2012). Choral singing, health and quality of life: The story of Diana. *Arts and health, 4*(3), 249–261.
Balsnes, A. H. (2014). "I get *sick* when I don't go to choir practice." Choral singing as a health promoting resource. In U. Geisler & K. Johansson (Eds.), *Choral singing: Histories and practices*. Cambridge: Cambridge Scholars Publishers.
Balsnes, A. H. (2017). The silver voices: A possible model for senior singing. *International Journal of Community Music, 10*(1), 59–69.
Balsnes, A. H., & Jansson, D. (2015). Unfreezing identities. Exploring choral singing in the workplace. *International Journal of Community Music, 8*(2), 163–178.
Balsnes, A. H., & Schuff, H. (2013). "Min styrke når jeg er svak på fremmed sted." Om korsang, flyktningehelse og integrering. In H. Schuff, H. R. Salvesen, & H. Hagelia (Eds.), *Forankring og fornyelse: Festskrift for Ansgarskolen 1913–2013*. Kristiansand: Portal forlag.
Batt-Rawden, K. (2010). The role of music in a salutogenic approach to health. *International Journal of Mental Health Promotion, 12*(2), 2–19.
Bjøru, A.M.D. (2015). *Korsang og psykisk helse. Empirisk studie om et tilrettelagt korsangtilbud for personer med psykiske helseplager.* Masteroppgave i psykisk helsearbeid, Høgskolen i Nord-Trøndelag.
Bonde, L. O. (2011). Health music(k)ing – Music therapy or music and health? A model, eight empirical examples and some personal reflections. *Music and Arts in Action*. Special Issue: Health promotion and wellness, *3*(2), 120–140.
Clift, S., Hancox, G., Morrison, I., Hess, B., Stewart, D., & Kreutz, G. (2008). *Choral singing, well-being and health, Sidney De Haan Reports*. Canterbury: Sidney De Haan Research Centre for Arts and Health.
Clift, S., Jennifer, N., Raisbeck, M., Whitmore, C., Morrison, I. (2010). Group singing, wellbeing and health: A systematic mapping of research evidence. *UNESCO Observatory, Faculty of Architecture, Building and Planning, The Melbourne Refereed E-journal, 2*(1), 1–25.
Dalgard, O. S., Thapa, S. B., Hauff, E., McCubbin, M., & Syed, H. R. (2006). Immigration, lack of control and psychological distress: Findings from the Oslo Health Study. *Scandinavian Journal of Psychology, 47*, 551–558. https://doi.org/10.1111/j.1467-9450.2006.00546.x.
DeNora, T. (2013). *Music asylums. Wellbeing through music in everyday life*. Surrey: Ashgate.
Fugelli, P. (1998). Folkehelse – folkets helse? *Tidsskrift for den Norske Lægeforening, 118*, 1421–1425.
Kreuz, G., & Brünger, P. (2012). A shade of grey: Negative associations with amateur choral singing. *Arts and Health, 4*(3), 230–238.
Kvale, S. (2007). *Doing interviews*. London: Sage. (The SAGE Qualitative research kit).
Ruud, E. (2010). *Music therapy: A perspective from the humanities*. Gilsum: Barcelona Publishers.
Ruud, E. (2013). *Musikk og identitet*. Oslo: Universitetsforlaget.
Skingley, A., Page, S., Clift, S., Morrison, I., Coulton, S., Treadwell, P., et al. (2013). "Singing for Breathing": Participants' perceptions of a group singing programme for people with COPD. *Arts & Health, 6*(1), 59–74.
Small, C. (1998). *Musicking. The meanings of performing and listening*. Hanover: University Press of New England. (Music/culture).
Stake, R. E. (2000). Qualitative case studies. In N. K. Denzin & Y. S. Lincoln (Eds.), *Handbook of qualitative research* (pp. 443–481). Thousand Oaks: Sage.
Stige, B. (2006). On a notion of participation in music therapy. *Nordic Journal of Music Therapy, 15*(1), 121–138.
Stige, B. (2012). Health musicking: A perspective on music and health as action and performance. In R. MacDonald, G. Kreutz, & L. Mitchell (Eds.), *Music, health and wellbeing* (pp. 183–195). Oxford: Oxford University Press.
Yin, R. K. (1989). Case study research. In *Design and methods*. Newbury Park: Sage.

Chapter 12
Humanizing the Working Environment in Health Care Through Music and Movement

The Importance of Embodied Leadership

Eva Bojner Horwitz

Introduction

Work-related stress is thought to be one of the main causes of burnout (Golkar et al. 2014; Nordlund et al. 2010; Zoni and Lucchini 2012; Åsberg et al. 2010). In Sweden today, a significant portion of health-care personnel report high levels of work-related stress, fatigue and mental disorders, including burnout (Grossi et al. 2015; Swedish Social Insurance Agency 2015). If health-care personnel are experiencing high levels of stress and burnout, this is likely to negatively impact the quality of care that they can provide and subsequently affect patient satisfaction (Hayes et al. 2010; Kutney-Lee et al. 2013; Shanafelt et al. 2015). Moreover, work-related stress is also responsible for costly sick leave, especially in the health-care sector (Moss et al. 2016). So far, there are few interventions in place that effectively prevent or ameliorate this worrying state of affairs, even though we know that employees, both men and women, who report lack of decision latitude and job strain will experience increase of depression symptoms (Theorell et al. 2015) and have an increased incidence of ischaemic heart disease over time (Theorell et al. 2016). It has long been suggested that organizational-level dynamics can be significant in causing stress, such as fear of job loss during restructuring, budget cuts, heavy workloads and lack of social support or control over one's work (Denton et al. 2002; Landsbergis 1988). The importance of high level of job support and workplace justice is a factor shown to be protective for emotional exhaustion and high workload (Aronsson et al. 2017). Therefore more specific organizational interventions targeting psychosocial factors are needed (Theorell et al. 2015).

E. B. Horwitz (✉)
Center for Social Sustainability, CSS, Karolinska Institutet, Stockholm, Sweden

Department of Clinical Neuroscience, Karolinska Institutet, Stockholm, Sweden

Department of Public Health and Caring Sciences, Uppsala University, Uppsala, Sweden
e-mail: Eva.Bojner.Horwitz@ki.se

L. O. Bonde, T. Theorell (eds.), *Music and Public Health*,
https://doi.org/10.1007/978-3-319-76240-1_12

Organizational leadership of course must play a key role in any attempt to reduce work-related stress. In using the concept of leadership, we do speak, on one hand, about those people in formal leadership roles within an organization, who have a responsibility to shape how workloads are distributed and job insecurity is minimized; but we also speak, on the other hand, about the development of workplace culture and a continual process of taking care of those within the organization and assuming a "responsibility for the wellbeing of organizations and people" (Romanowska et al. 2016). There has been a growing interest in how an empathetic environment can be of great importance for successful leadership: both empathy towards oneself and empathy towards others (Bojner Horwitz and Huss 2016). It can be easy to take for granted that the health-care leaders have inner qualities that enable them to understand themselves and others. Instead, it may be that further knowledge of how to develop such empathy may be needed. With a focus on music, this chapter explores the way in which cultural activities can help to foster empathy, discussing how health care can be "humanized". And whilst the concept of such humanity is complex in its philosophical-ethical foundations and embeddedness in social relations, as illustrated in some of the discussions around developing humane health systems for patients (Pijnenburg 2002), it is still possible and important to consider how working with individual health-care workers can contribute to transforming health care into a system where its participants are gentler, kinder or more trusting towards each other.

Looking at the humanization through the lens of empathy, it is useful to consider where and why empathy can be lacking. The concept of alexithymia was introduced by Sifneos (1973) who observed that certain people had difficulties in recognizing emotions and finding appropriate words to label them (see Chap. 4 by Theorell and Ullén). In Greek, the word *alexithymia* means "no words for feelings". The prevalence of alexithymia has been shown to be 5–10% for women and 9–17% for men in a Finnish working age population (Mattila et al. 2007). Alexithymia is related to an impairment of both emotional awareness and emotional regulation (Taylor et al. 1997). An individual who can operationalize and symbolize emotional responses, and who can regulate emotions, is said to be "less alexithymic" (Thompson 1994). Thinking about the dynamics of workplace in general, and the demands of leadership in particular, the possibility of alexithymia is one that it is important to be aware of because if we cannot recognize and regulate our own emotions, it is unlikely that we achieve this in our relations with others. In the long term, a lack, or insufficient level, of empathetic capability is problematic beyond the individual and could affect the community of the organization since the suppression and noncommunication of feelings (especially negative feelings) can trigger defensive behaviours and stress reactions. Although there are few studies on empathy and alexithymia, we have seen from twin studies that achievement in creative practices seems to be connected to a lower level of alexithymia (Lennartsson et al. 2017). It is this connection that leads us to explore how participation in creative practices may help to improve ability to empathize with others.

The arts have long been used by humans to express the incomprehensible: creative practices have been tools with which we have explored different levels of our consciousness and even to transform our experiences of life. Music and other forms of art used to be integral to our daily lives: our songs, rituals, paintings, sculptures and dances were an essential part of communicating, remembering, community formation, food production and hunting (Sonke-Hendersen 2007). Artistic expressions have been important for maintaining social order within institutions – norms, rules and codes of conduct – and cultural activities such as games were a form of creative problem-solving, helping tribes, villages and families to find solutions to new challenges and to improve the quality of their lives. In health care, music has been used as a form of co-therapy (Bonde 2007), in particular for the treatment of psychiatric disorders and of children (Covington 2001; Aldridge et al. 1995), but less attention has been paid to how it might contribute to ameliorating the experience of health care for members of staff. We seek to understand how music and other artistic practices relate to the development of emotional awareness and how this might be harnessed to foster empathy in health-care leaders and empathetic relations among staff and between staff and patients in a health-care setting. Of particular interest is the relationship between the physical and mental processes that are involved in undertaking music and other arts-related activities.

Embodying Empathy

Embodied Knowledge and Cognition

Scientists have discussed knowledge as being grounded in the body as "embodied". Although the term embodiment has been used in various ways (Niedenthal et al. 2005), definitions of embodiment share a theoretical focus on the brain's modality-specific systems that are constituted by:

1. The sensory system, which regulates perception of a current situation
2. The motor systems, which make action possible
3. The introspective systems, which govern cognitive operations and conscious experiences of emotion

These systems allow us to describe the ways in which cognition is embodied or, in other words, has its basis in the physical. This means that the way we perceive and make sense of the world is a function not only of our brain's cognitive function but also of our sensory and motor functions as well. There are likely no socially meaningful processes without a physical component that are not embodied (Niedenthal et al. 2005).

This framework is helpful in conceptualizing "empathy". Empathy can be defined in many ways, but emotional identification with another person is usually included as a criterion. Empathy is not only the ability to understand another person's emotional state but is also a process of understanding another person's feelings *in* ourselves (De Vignemont and Singer 2006). This implies that empathy is manifest at least partly as a physical sensation, a bodily or "embodied" identification of another person's emotional state. As Russo describes, "Empathy is a visceral and cognitive understanding of another's emotions or motivations" (McGarry and Russo 2011; Gallese 2005; Rizzolatti and Arbib 1998). Stern describes empathy as an "embodied affective resonance" which suggests that, together with the cognitive processes, the body-mind is always involved in the empathy response (Stern 1985/2000).

Mirror Neuron Activity

It is common refrain that someone else's mood, happy or sad, can "rub off" on others. Even in the eighteenth century, moral philosophers recognized the phenomenon of emotional mirroring; observing another (person) expressing emotion can evoke not only sympathy for their situation but also a visceral emotional identification. Adam Smith is often credited with describing this phenomenon as a "mirroring" effect, whereby we imagine ourselves in another person's situation and feel what she feels (Smith 1759). Smith argued that this ability for empathy (though he described it as sympathy) is an innately human quality that is essential for the maintenance of social order. Indeed, the capacity for empathy is, according to Smith, the bedrock of human society.

Smith's theories regarding mirroring were generally ignored (except for William James in the late nineteenth century, along with Wolfgang Prinz and others in the 1980s) until recently when scientists began to uncover neurophysiological evidence of changes in brain activity as a result of observing another person's emotions or movements. In the mid-1990s, a group of Italian neuroscientists working with non-human primates made a discovery of great significance for our understanding of interpersonal interaction (Gallese et al. 1996). They studied the brain activity in monkeys as they performed simple reaching and grasping movements and found that a monkey's hand motions activated not only the monkey's own sensorimotor neurons but also similar neurons within the brain of the monkey who had witnessed the movement (Umilta et al. 2001). Some years later, the researchers used the knowledge from the discovery of the innate mirroring processes at play within sensorimotor brain regions to try to explain the social, kinaesthetic and emotional reading abilities of humans (Gallese 2005). When we observe gestures, facial expressions and other movements, evidence suggests that some of the same parts of our brains are activated as those of the person that we watched move. Perhaps surprisingly, this suggests that there is an automated physical dimension to our response to others. Some neurophysiological work has to be done to reliably demonstrate that there are specific mirror neurons per se, in the human brain. With the help of an fMRI camera, researchers have illustrated how mirror neuron-like systems are linked to various human reactions (Wicker et al. 2003). For example, when subjects were asked to watch a person's different facial expressions when he smelled something pleasurable and then smelled something foul, the subjects were found to have the same emotional response, in the same parts of the brain (amygdala), as the person who experienced the scents.

So, how does this sensorimotor mirroring relate to empathy, what we might describe as emotional mirroring? Mirror neuron activation always takes place in relation to someone else; it is so-called "relational-activated". Ekman and Friesen (2003), saw facial expressions as something universal that affect our emotional reading of and identification with another person. People who automatically respond with a greater number of facial expressions in social contexts, who have higher "mimic capacity", also score higher on empathy scales (Chartrand and Bargh 1999). This is the reason why therapists helping people with reduced mimic capacity, such as those suffering from autism, increasingly use movement and facial expression exercises to activate empathy. Approaching emotion and empathy regulation via facial expressions may increase the activity of mirror neurons and their link to the limbic system (Chartrand and Bargh 1999). The study shows that people who rate highly for empathy also exhibit high mirror neuron activity in their emotional brain. There are of course a population of people, for example, those diagnosed with psychopathy, who may have learned which facial expressions are expected in certain social situations but lack the feelings behind them. Psychopaths have indeed been found to have a reduced capacity for empathy as well as dysfunction in the mirror neuron system when observing others expressing emotions (McGarry and Russo 2011).

It follows to consider whether our automated physical response to others be harnessed in other ways to foster empathy: some evidence suggests that the mirroring of another person's movements can be used to access their emotions. Mirror neurons in non-human primates have been located near areas in the human homologue brain that support not only visual an audio sensory information but also tactile sensations. Stern's understanding of empathy suggests that mirror neurons are linked to both the prefrontal cortex (the part of the human brain associated with higher cognitive functions and reasoning) and the limbic system (the network of brain structures associated with emotional processing) (Stern 1985/2000). This means that a mirrored motion may also provide a way of manifesting the mirrored person's emotions within oneself.

Training Emotion and Empathy

Research into the functioning of mirror system neurons indicates that there are possibilities for training our emotional repertoire and building empathy. Visual observations that can be used in these exercises include body movement and/or various facial expressions such as fear, joy, disgust and so on (Rizzolatti and Arbib 1998). Some mirror neurons are activated because of previous experience to which we can refer. This means that in an empathetic event, it may be important that we have former experience to provide a reference that allows the process of mirroring to be activated. Researchers therefore believe that it is possible to "train" mirror system neurons to activate in a way that will promote empathetic responses to others (Gallese 2005; Stern 1985/2000). We might argue that an awareness of how our own knowledge is shaped by sensations, and a greater connection to our inner world through practice of mirroring, is important in fostering empathy, which seeks to expand our awareness to incorporate others.

A group of researchers in Toronto (Russo et al 2013) have created an "emotional bank" of different emotions expressed through movements, facial expressions and gestures. The emotional bank can be used to train individuals who have difficulties reading both their own emotions and those of others. It uses films of actors who perform emotions. The films are used therapeutically to train and develop individual capacity for emotional differentiation, which means that a certain emotion can be expressed by more or less explicitly through mimicry.

In the same way, as in facial expression experiments, movements in an emotional bank are powerful tools for helping individuals improve their capacity for empathy towards themselves and others. One reason for this is that the mimicry of movements may contribute to an embodied sense of belonging. This sense of belonging is also linked to the mirror neuron activity in our brains that occurs after watching our own movements mimicked (mirrored) by another person (Berrol 2006). Mirroring exercises have also been developed from Berrol's research to promote capacity for empathy: the exercise involves exaggerated sequences of movement that are linked to different emotional expressions, much like in the emotional bank. In the mirror exercise, patients are asked to gradually scale down the movements that they are mimicking so that they become more natural. This training has been used with stroke patients with aphasia who have difficulties expressing or interpreting speech (Schlaug et al. 2008).

Cultivating Empathy Through Cultural Activities

Think of all the times you have experienced an emotional response to a song, work of art or ballet, without really being able to articulate why it affected you so deeply. Engaging with cultural activities allows us to "sneak through the back door into the brain", reaching the emotional brain regions without first passing through the brain structures involved with rational or cognitive processes (such as the prefrontal cortex): Le Doux (1998) calls this the "surprise effect" of cultural activities. Le Doux argues that some responses go on a "fast track" directly from the thalamus down to the amygdala nuclei without first passing through parts of the brain associated with rational thought. We become affected by stimuli without engaging in a rational process to understand why. In the same way, body movements – seen or experienced – can invoke emotional responses, reaching parts of the limbic system via sensorimotor neurons, thus bypassing cognitive systems. Music, integrated with two related forms of cultural activity – dance and theatre – was used as part of an experiment designed to promote empathy to people working in a health-care setting, *The Cultural Palette*. Here I discuss how music plays a vital role in facilitating an engagement in mirroring through the use of cultural activities.

Our focus theoretically so far has been on movement and its potential to access and develop an embodied empathetic response, but music can play a key role in enhancing this process. It is known that areas within the brain that involve rhythm and pulse perception overlap areas that control our movements in group of subcortical nuclei termed the basal ganglia (Grahn and Brett 2007). This means that music has the potential to enhance empathetic movements; in particular, the rhythm of music can facilitate the reading of another person's movement pattern. By dancing, we can improve our ability to become more aware of our own emotions and the emotions of others: the more we dance, the more we train our mirror systems and their links to the limbic system (Calvo-Merino et al. 2005). By watching someone else dance, our sensorimotor brain regions are activated (Bläsing et al. 2012). Interestingly, with regard to reading other's emotions, we have seen in a large sample of the Swedish population that engagement in dance is associated with emotional competence in interaction with others (Bojner Horwitz et al. 2015). The result from this cohort (n = 5431) corresponds to a more developed awareness of emotional processing and higher ability to interpret the emotions of others in dancers than in nondancers. In other research, we have also explored the effects of non-participative experience of movement using theatre. We asked patients who were experiencing pain to act in the play *Medea* and then to watch the same piece performed by professional actors (Bojner Horwitz et al. 2010). Patients were found to react more strongly to watching the play (passive consumption) than to acting out the play themselves (active participation). It appears the patients in the study were more alert when passively viewing and could more easily identify with the actor, which fostered a stronger reaction than when they were preoccupied with acting out the piece themselves.

Although this area of research is still in its early stages, there is increasing evidence that we have sophisticated mechanisms for the "emotional reading" of movements and that we can even improve this capacity through training (McGarry and Russo 2011; Cross et al. 2009). Music plays a crucial role in artistic expressions that involve movement, such as dance. Try watching a dance video with the sound off. Does it seem flatter, less evocative? We all know that music can on its own arouse strong feelings, but combining it with dance can have a reinforcing function, enhancing our emotional experience. This phenomenon implies that a person who has difficulty reading emotions in a dance alone may be better able to understand its gestures when accompanied by music (Bojner Horwitz 2004).

Music and dance used in combination also strengthen what we call multimodality in perception of an experience. Multimodality is when different modalities such as movement, vision, somatosensation or sound are experienced at the same time. Recent evidence from brain scans suggests that the greater number of modalities through which a novice dancer has learned a new dance sequence, the greater engagement of sensorimotor brain regions is seen when the dancer watches those movements back (Kirsch and Cross 2015). It has also been suggested that music enhances our emotional response to observed movements (Moreno and Mayer 2007). This suggests that music forms a significant part of the library of emotional references that we build up and which help us to empathize with others.

Cultural Activities and Modalities of Consciousness

Different individuals are likely to have different code systems to access this "back door" to the emotional brain, and so it is important to find a way to evaluate which stimuli work best for which individuals. By being aware of these preferences, we find ourselves in a better position to select an appropriate empathetic artistic activity for use in empathy training. According to previous research, it appears that different cultural activities can code different levels of our conscious mind (Bojner Horwitz 2011), but this does not necessarily conflict with the bypassing of the rational mind as described in the process above: George Downing (1996) categorized conscious awareness into different modalities of consciousness, physical, emotional, sensory, visual and intellectual. He argued that, because consciousness is not unitary, an understanding of the modality of consciousness that we are using can help increase our understanding of the possibilities of empathetic coding, the code for how we can access our emotional brain via the most direct route. According to Downing, increasing awareness itself is therefore an essential part of training – if we can train ourselves to open up as many modalities of consciousness as possible, we can broaden our susceptibility to empathy, what we might think of as our empathetic repertoire.

Implementing Music and Movement in Swedish Health Care

In Sweden, we have implemented and used the knowledge from the different areas of theory presented above, mixed modalities, mirror system, emotional brain, embodiment and emotional reading, in three research projects that sought to transform the working environment of the health-care system through engagement with cultural activities.

The first project was the *Cultural Palette* (Grape Viding et al. 2015): a randomized controlled study, wherein we demonstrated that cultural activities have a positive effect in treating symptoms of burnout in female patients. Six different activities (theatre, movie, singing, dancing, mindfulness, and taking part in a musical show) were offered once a week, lasting 90 min per session. The results showed decreased levels of exhaustion, increased ability to express emotions (decrease in alexithymia) and increased levels of self-rated health in the patient group. After four separate interventions at four different health-care centres, focus group interviews were also conducted (Grape et al. 2017). From the focus groups, we observed a ripple effect of both trust and empathy within both the patients and the health-care staff, even though members of staff were not actively participating in the palette of activities.

A second study targeted staff specifically – the *Cultural Palette for staff* (Bojner Horwitz et al. 2017) – which incorporated music, also sought to promote empathy through embodiment and mirroring emotions. In this study, the impact of self-chosen arts-based activities such as qigong, yoga, line-dancing and baking was evaluated over 10 weeks in three different health-care centres. The study's findings showed that the arts-based activities (a) helped to reduce individual stress and to enhance mood over time, (b) helped to transform workplace relationships within wards and (c) humanized the overall work-climate in the health-care setting. There were also challenging tasks for the staff, for example, difficulties in breaking the habituated patterns of behaviour that were related to their stress. Overall, we found that the arts-based activities introduced to the health-care workers were transformative of the individual, group and work culture. We argued that these transformations, which promoted greater care among and for the staff, are also a step towards providing better care for patients.

Finally, we conducted the *Playmäkers project* in Södertälje (Bojner Horwitz and Huss 2016). The *Playmäkers* project was a multipurpose, arts-based intervention for young adults. Young people were asked to interact with members of the elderly population who have dementia. The project aimed to harness the technological sophistication of young adults (who were named *Playmäkers*) to offer elderly people suffering from dementia a way of accessing music (involving movies and other visual arts) reminiscent of their own youth. *Playmäkers* used iPads to access these cultural products, and the technology served to mediate communication between members of the two generations. The process also included nurses and patients' relatives. Interactive outcomes of the *Playmäkers* project were analysed with qualitative data and were theorized with regard to non-verbal and embodied communication, the emotional brain and the role of the arts in health. The project found that both dementia sufferers and their relatives were emotionally stimulated and that the activities provided a mediated space within which to connect to others. For those health-care workers tasked with caring for dementia patients, the activities were found to humanize the heavy workload of caring. For the *Playmäkers*, the intervention helped to mitigate a culture that is heavily reliant on technology and technologically mediated communication, by facilitating an experience that involves technology *and* human interaction, empathy and emotional intelligence. Despite these positive results, an important finding with regard to the implementation of such interventions in health care is that, for some health-care workers, the *Playmäker* activity was also experienced as a threat to their role, as something that prevented them from meaningful emotional interactions with their patients. Some also felt that, although the activities were meaningful, they had no time to participate in them because of their heavy workload, echoing concerns about the organizational-level contributors to stress in health-care workers (see Denton et al. 2002) and emphasizing how such organizational dynamics could inhibit interventions at the level of the individual.

Future Directions for Empathy and the Arts in Health Care

In modern workplaces stress and burnout are pressing issues to be considered by organizational leaders, no context more so than that of health care in which the safety of patients is at stake. In the face of financial pressures, restructuring of health-care organizations and the challenge of already demanding workloads, which have long been thought to contribute to stress (Denton et al. 2002; Landsbergis 1988), it will be essential that any interventions are able to tackle a given problem on multiple levels, such as that of the *Cultural Palette*, and to recognize where organizational structures and social relations may inhibit the potential positive effects of individual-level interventions such as in the *Playmäkers* project. By seeking to contribute to easing the problem of stress and burnout at the level of the individual through the development of empathy, it becomes possible to also contribute to creating a working environment that is not only less stressful and more satisfying for both staff and patients but which is also permeated with a culture of care and is more *human*. Staff and patient care need to go hand in hand – in the words of one of the patients involved in our *Cultural Palette* study, "you can't feel better than your caregiver". In this way, empathy becomes a way of practising leadership through taking responsibility for the wellbeing of those in the organization. The task is to find a way to transform our health-care centres and hospitals holistically as an organism of intertwined stakeholders who are not abstracted objects but subjects with feelings and needs. In seeking to find ways to promote empathy, we have considered both the embodied experience of emotion and the embodied reaction that we have to the emotions of others (Crosta Ahlforn et al. 2017). By harnessing the powerful emotional interaction between music and movement, we have built programmes of cultural activities to explore how the body can be used to (a) increase awareness of one's own emotions and (b) increase one's understanding of the experiences of others. As evidence builds for the ways in which engagement with artistic programmes offers value to health care, it is important that we consider the variations in the experience of these programmes between individuals, staff, groups and types of organization, how such interventions sit within an interplay of individual and organizational-level dynamics and how programmes such as those described can be best implemented with the use of combined resources.

References

Aldridge, D., Gustorff, D., & Neugebauer, L. (1995). A preliminary study of creative music therapy in the treatment of children with developmental delay. *The Arts in Psychotherapy, 22*(3), 189–205.

Aronsson, G., Theorell, T., Grape, T., Hammarström, A., Hogstedt, C., Marteinsdottir, I., et al. (2017). A systematic review including meta-analysis of work environment and burnout symptoms. *BMC Public Health, 17*, 264.

Åsberg, M., Krakau, I., Nygren, Å., Rodhe, M., Wahlberg, A., & Währborg, P. (2010). Stress som orsak till psykisk ohälsa /stress as the cause of mental illness. *Läkartidningen, 107*(19–20), 1307–1310.

Berrol, C. F. (2006). Neuroscience meets dance/movement therapy: Mirror neurons, the therapeutic process and empathy. *The Arts in Psychotherapy, 33*(4), 302–315.

Bläsing, B., Calvo-Merino, B., Cross, E. S., Jola, C., Honisch, J., & Stevens, C. J. (2012). Neurocognitive control in dance perception and performance. *Acta Psychologica, 139*(2), 300–308.

Bojner Horwitz, E. (2004). Dance/movement therapy in fibromyalgia patients: Aspects and consequences of verbal, visual and hormonal analyses (Doctoral dissertation, Acta Universitatis Upsaliensis).

Bojner Horwitz, E. (2011). *Kultur för hälsans skull. (Culture for the sake of health).* Stockholm: Gothia Förlag AB.

Bojner Horwitz, E., & Huss, E. (2016). Using internet based arts to promote inter-generational meetings between young people and senior citizens: The Playmäkers project in Sweden. *Journal of Applied Arts & Health, 7*(3), 297–311.

Bojner Horwitz, E., Kowalski, J., & Anderberg, U. M. (2010). Theatre for, by and with fibromyalgia patients – evaluation of emotional expression using video interpretation. *The Arts in Psychotherapy, 37*(1), 13–19.

Bojner Horwitz, E., Lennartsson, A.-K., Theorell, T., & Ullén, F. (2015). Engagement in dance is associated with emotional competence in interplay with others. *Frontiers in Psychology, 31*, 1–8.

Bojner Horwitz, E., Grape Viding, C., Rydwik, E., & Huss, E. (2017). Arts as an ecological method to enhance quality of work experience of healthcare workers: A phenomenological hermeneutic study. *Journal of Qualitative Studies in Health and Well-being,* Vol 13, 1333898.

Bonde, L. O. (2007). Music as co-therapist: Investigations and reflections on the relationship between music and imagery in the bonny method of guided imagery and music. In I. Frohne-Hagemann (Ed.), *Receptive music therapy* (pp. 43–74). Wiesbaden: Dr. Ludwig Reichert Verlag.

Calvo-Merino, B., Glaser, D. E., Grèzes, J., Passingham, R. E., & Haggard, P. (2005). Action observation and acquired motor skills: An FMRI study with expert dancers. *Cerebral Cortex, 15*(8), 1243–1249.

Chartrand, T. L., & Bargh, J. A. (1999). The chameleon effect: The perception –behaviour link and social interaction. *Journal of Personality and Social Psychology, 76*(6), 893.

Covington, H. (2001). Therapeutic music for patients with psychiatric disorders. *Holistic Nursing Practice, 15*(2), 59–69.

Cross, E. S., Kraemer, D. J., Hamilton, A. F. D. C., Kelley, W. M., & Grafton, S. T. (2009). Sensitivity of the action observation network to physical and observational learning. *Cerebral Cortex, 19*(2), 315–326.

Crosta Ahlforn, K., Bojner Horwitz, E., & Osika, W. (2017). A Swedish version of the Consultation and Relational Empathy (CARE) measure. *Scand J of Primary Health Care, 35*(3), 286–92.

De Vignemont, F., & Singer, T. (2006). The empathic brain: How, when and why? *Trends in Cognitive Sciences, 10*(10), 435–441.

Denton, M., Zeytinoglu, I. U., Davies, S., & Lian, J. (2002). Job stress and job dissatisfaction of home care workers in the context of health care restructuring. *International Journal of Health Services, 32*(2), 327–357.

Downing, G. (1996). *Kroppen och ordet. Kroppsorienterad psykoterapi—teoretisk bakgrund och klinisk tillämpning (The body and the word. Bodily oriented psychotherapy—theoretical background and clinical application).* Borås: Natur och Kultur.

Ekman, P., & Friesen, W. V. (2003). *Unmasking the face: A guide to recognizing emotions from facial clues.* Los Angeles: Malor Books.

Frank A. Russo, Naresh N. Vempala, Gillian M. Sandstrom, (2013) Predicting musically induced emotions from physiological inputs: linear and neural network models. Frontiers in Psychology 4.

Gallese, V. (2005). "Being like me": Self-other identity, mirror neurons and empathy. *Perspectives on Imitation: From Cognitive Neuroscience to Social Science, 1*, 101–118.

Gallese, V., Fadiga, L., Fogassi, L., & Rizzolatti, G. (1996). Action recognition in the premotor cortex. *Brain, 119*(2), 593–609.

Golkar, A., Johansson, E., Kasahara, M., Osika, W., Perski, A., & Savic, I. (2014). The influence of work-related chronic stress on the regulation of emotion and on functional connectivity in the brain. *PLoS One, 9*(9), e104550.

Grahn, J. A., & Brett, M. (2007). Rhythm and beat perception in motor areas of the brain. *Journal of Cognitive Neuroscience, 19*(5), 893–906.

Grape Viding, C., Osika, W., Theorell, T., Kowalski, J., Hallqvist, J., & Bojner Horwitz, E. (2015). "The culture palette" – a randomized intervention study for women with burnout symptoms in Sweden. *British Journal of Medical Practitioners, 8*(2), a813.

Grape Viding, C., Osika, W., & Horwitz, B. (2017). "You can't feel healthier than your care-giver" – the ripple effect of trust and empathy for patients and health care staff, cultivated through cultural activities. *The Journal of Nursing Care, 6*(5), 1–8.

Grossi, G., Perski, A., Osika, W., & Savic, I. (2015). Stress-related exhaustion disorder–clinical manifestation of burnout? A review of assessment methods, sleep impairments, cognitive disturbances, and neuro-biological and physiological changes in clinical burnout. *Scandinavian Journal of Psychology, 56*(6), 626–636.

Hayes, B., Bonner, A., & Pryor, J. (2010). Factors contributing to nurse job satisfaction in the acute hospital setting: A review of recent literature. *Journal of Nursing Management, 18*, 804–814.

Kirsch, L. P., & Cross, E. S. (2015). Additive routes to action learning: Layering experience shapes engagement of the action observation network. *Cerebral Cortex, 25*(12), 4799–4811.

Kutney-Lee, A., Wu, E. S., Sloane, D. M., & Aiken, L. H. (2013). Changes in hospital nurse work environments and nurse job outcomes: An analysis of panel data. *International Journal of Nursing Studies, 50*(2), 195–201.

Landsbergis, P. A. (1988). Occupational stress among health care workers: A test of the job demands-control model. *Journal of Organizational Behavior, 9*(3), 217–239.

Le Doux, J. E. (1998). *The emotional brain: The mysterious underpinnings of emotional life*. New York: Weidenfeld & Nicolson.

Lennartsson, A. K., Bojner, H. E., Theorell, T., & Ullén, F. (2017). Creative artistic achievement is related to lower levels of alexithymia. *Creativity Research Journal, 29*(1), 29–36.

Mattila, A. K., Ahola, K., Honkonen, T., Salminen, J. K., Huhtala, H., & Joukamaa, M. (2007). Alexithymia and occupational burnout are strongly associated in working population. *Journal of Psychosomatic Research, 62*(6), 657–665.

McGarry, L. M., & Russo, F. A. (2011). Mirroring in dance/movement therapy: Potential mechanisms behind empathy enhancement. *The Arts in Psychotherapy, 38*(3), 178–184.

Moreno, R., & Mayer, R. (2007). Interactive multimodal learning environments. *Educational Psychology Review, 19*(3), 309–326.

Moss, M., Good, V. S., Gozal, D., Kleinpell, R., & Sessler, C. N. (2016). A critical care societies collaborative statement: Burnout syndrome in critical care health-care professionals. A call for action. *American Journal of Respiratory and Critical Care Medicine, 194*(1), 106–113.

Niedenthal, P., Barsalou, L., Winkielman, P., Krauth-Gruber, S., & Ric, F. (2005). Embodiment in attitudes, social perception and emotion. *Personality and Social Psychology Review, 9*(3), 184–211.

Nordlund, S., Reuterwall, C., Höög, J., Lindahl, B., Janlert, U., & Slunga Birgander, L. (2010). Burnout, working conditions and gender – results from the northern Sweden MONICA study. *BMC Public Health, 10*(1), 326–334.

Pijnenburg, M. (2002). Humane healthcare as a theme for social ethics. *Medicine, Health Care and Philosophy, 5*(3), 245–252.

Rizzolatti, G., & Arbib, M. (1998). Language within our grasp. *Trends in Neurosciences, 21*, 188–194.

Romanowska, J., Nyberg, A., & Theorell, T. (2016). *Developing leadership and employee health through the arts: Improving leader-employee relationships*. Cham: Springer Books.

Schlaug, G., Marchina, S., & Norton, A. (2008). From singing to speaking: Why singing may lead to recovery of expressive language function in patients with Broca's aphasia. *Music perception: An interdisciplinary journal, 25*(4), 315–323.

Shanafelt, T. D., Gorringe, G., Menaker, R., Storz, K. A., Reeves, D., Buskirk, S. J., et al. (2015). Impact of organizational leadership on physician burnout and satisfaction. *Mayo Clinic Proceedings, 90*(4), 432–440.

Sifneos, P. E. (1973). The prevalence of 'alexithymic' characteristics in psychosomatic patients. *Psychotherapy and Psychosomatics, 22*(2), 255–262.

Smith, A. (1759). *The theory of moral sentiments* (Vol. 1976, p. 238). Oxford: Clarendon Press. Accessed via http://file.org.uk/ebooks/65/7.pdf 05/05/2017.

Sonke-Hendersen, J. (2007). History of the arts and health across cultures. In Irlene A Serline (Ed.) *Whole person health care* Vol 3. The Arts in Health. Praeger: Santa Barbara, CA. ISBN 13:9780275992347.

Stern, D. N. (2000/1985). *Interpersonal world of the infant: A view from psychoanalysis and development psychology*. New York: Basic Books.

Swedish Social Insurance Agency. (2015). Sjukskrivningar 60 dagar eller längre. En beskrivning av sjukskrivna åren 1999–2014 efter kön, ålder, arbetsmarknadsstatus, yrke, sjukskrivningslängd och diagnospanorama. Report no. 2015:1. Stockholm: Försäkringskassan.

Taylor, G. J., Bagby, R. M., & Parker, J. D. A. (1997). *Disorders of affect regulation: Alexithymia in medical and psychiatric illness*. Cambridge: Cambridge University Press.

Theorell, T., Hammarström, A., Aronsson, G., Träskman Bendz, L., Grape, T., Hogstedt, C., et al. (2015). A systematic review including meta-analyses of work environment and depressive symptoms. *BMC Public Health, 15*, 738.

Theorell, T., Jood, K., Slunga Järvholm, L., Vingård, E., Perk, J., Östergren, P. O., et al. (2016). A systematic review of studies in the contributions of the work environment to ischaemic heart disease development. *The European Journal of Public Health, 26*(3), 470–477.

Thompson, R. A. (1994). Emotional regulation: A theme in search of definition. In N. A. Fox (Ed.), *The development of emotion regulation: Biological and behavioural considerations* (pp. 25–52). Chicago: The University of Chicago Press. (Monographs of the Society for Research in Child Development, 59, Serial No. 240).

Umilta, M. A., Kohler, E., Gallese, V., Fogassi, L., Fadiga, L., Keysers, C., et al. (2001). I know what you are doing: A neurophysiological study. *Neuron, 31*(1), 155–165.

Wicker, B., Keysers, C., Plailly, J., Royet, J. P., Gallese, V., & Rizzolatti, G. (2003). Both of us disgusted in My insula: The common neural basis of seeing and feeling disgust. *Neuron, 40*(3), 655–664.

Zoni, S., & Lucchini, R. G. (2012). European approaches to work-related stress: A critical review on risk evaluation. *Safety and Health Work, 3*, 43–49.

Chapter 13
Music as an Integral Part of "Culture on Prescription"

Stine Lindahl Jacobsen, Helle Nystrup Lund, and Lars Rye Bertelsen

Introduction

In this article, we will focus on the recent development within the field of arts and health in Denmark and further zoom in on the use of music in a specific "arts on prescription" model carried out in Aalborg Municipality. Having offered a brief overview of established arts and health strategies in the rest of Scandinavia and the UK, we will next turn to the Danish project called "Culture Vitamins" and its use of music and music intervention. Following this, we will discuss political issues, applicability, and future research. Before we start, it is important to clarify that "music intervention" draws on two different fields from clinical practice including music medicine and music therapy. Music medicine is mainly a stimulus-response-focused model, whereas music therapy is a psychodynamic-humanistic model focused on interpersonal communication (Lund et al. 2016).

Arts and Health in the UK

In the last 20 years, "arts on prescription" and community-based arts and health interventions have been used, acknowledged, and supported in the UK. The impact of policy papers directly relating to arts in health has been significant for the field,

S. L. Jacobsen (✉)
Department of Communication and Psychology, Aalborg University, Aalborg, Denmark
e-mail: slj@hum.aau.dk

H. N. Lund
Aalborg University Hospital, Psychiatry, Aalborg, Denmark
e-mail: hnl@rn.dk

L. R. Bertelsen
Aalborg University & Aalborg University Hospital Psychiatry, Aalborg, Denmark
e-mail: larb@hum.aau.dk

L. O. Bonde, T. Theorell (eds.), *Music and Public Health*,
https://doi.org/10.1007/978-3-319-76240-1_13

including leading to increased funding for research and practice as well as better communication between the field of arts and the field of health. Strategic opportunities for the profiling of arts in health activity have led to increased awareness of how the arts can support health (Fancourt 2017). As many arts in health programs operate at more local levels and do not directly depend on government or other national endorsement to attract funding, partners, or participants, this political activity is often not well known among the general public. However, for programs seeking to expand from working at a local level to operating on a larger scale, such policy activity becomes of real significance in providing evidence of high-level support and endorsement for the field. Several arts and health organizations and movements have been established for years such as "Music and Health," but no government-funded large-scale research studies have been conducted. Because of the usual complex setting of community-based activities, the challenge seems to be to develop large-scale rigorous controlled effect studies and provide evidence to sustain good projects and collaborate across different arts- and health-related projects (Hamilton et al. 2015).

The inquiry report *Creative Health: The Arts for Health and Wellbeing* (2017) offers an extensive overview of the research on arts and health as there is quite an impressive research database on arts and health originating from the UK of which only a selected few is mentioned here. Fancourt (2017) lists both British and international art and health research under the following categories: critical care and emergency medicine, dentistry, geriatric medicine, healthcare staff, neurology, obstetrics, gynecology, neonatology, oncology, pediatrics, palliative care, public health, psychiatry, rehabilitation medicine, and surgery. Within public health one of the key research findings includes that attending cultural events is linked with better perceived health, and within psychiatry group drumming can reduce anxiety, depression, and inflammatory immune response in mental health service users, while dancing classes have also been shown to reduce anxiety and depression (Fancourt 2017). Furthermore, a comprehensive anthology on how music can positively affect health and well-being was co-authored by a Scottish music psychology professor in 2012, and here international research from many disciplines is presented and discussed. MacDonald points out how various countries have instigated national bodies to investigate the relationship between the arts and healthcare, for example, the Arts and Health (Australia), the Society for the Arts in Healthcare (USA), and Artful (Scotland) (MacDonald et al. 2012).

Arts and Health in Scandinavia

While culture and health projects gather momentum in the rest of Scandinavia, it appears that Denmark has been falling behind. According to Jensen and Wille (2016), the sociopolitical situation in Denmark requires a more interdisciplinary anchored approach where initiatives can take place cross-sector. To uncover unexplored potential in arts and health projects and to aid an interdisciplinary political approach, experts recommend for a national and a Scandinavian network to develop (Jensen and Wille 2016). The *Nordic Network for Research in Music and Public Health* – the network behind the present volume – is an example of how such interdisciplinary work can be organized (see Chap. 1 by Bonde and Theorell). In the following, examples of arts and health projects from the Scandinavian countries are presented aiming to offer a brief overview.

In Sweden, arts and health have been supported and acknowledged politically and financially as an important ingredient for public health since 2000. In 2007, the Swedish Parliament established the Culture and Health Association, and many large-scale projects, publications, and movements around arts and health have since been developed including "Kulturhälsoeboxen" and "Centrum for Kultur och Hälsa" (see Chap. 12 by Bojner Horwitz). Many different populations and aims are included such as increased participation by the elderly and rehabilitation for common health problems. Research into the effects and properties of arts and health is conducted by several universities, including the University of Gothenburg, Chalmers University of Technology, and Karolinska Institute in Stockholm. Promising results include increased quality of life, improved self-rated health, and decreased burn-out symptoms and mental health symptoms such as anxiety and depression (Jensen et al. 2016a). A recent large-scale "arts on prescription" study in Skåne with 123 participants examined the effect of participation in cultural activities for individuals with mental health problems and musculoskeletal pain. The study had no control group but showed promising results on improved health-related quality of life, function, and work ability also in a 12-month follow-up (Stigmar et al. 2016).

In Norway, both the Ministry of Culture and the Ministry of Health and Care Services support and acknowledge that participation in arts and cultural activities can have health-promoting effects, which is also emphasized in both public health law and cultural law. Since the early 1990s, the government has supported small-scale arts and health projects, and the field of practice is increasing where participants with various health problems engage in arts, culture, and nature activities. Research into the effects and properties of arts and health issues is performed by several agents including the Center for Research in Music and Health (CREMAH in Oslo), the Grieg Academy Music Therapy Research Centre (GAMUT in Bergen), and the Norwegian Resource Centre for Arts and Health (Volda University College). A significant part of arts and health research in Norway relates to music and health and music therapy. Already in 2006 publications had a focus on music and health, and today the literature is rich (Aasgaard, 2006). A focus on nature-culture interplay is also present in Norwegian research (Batt-Rawden and Tellnes 2012). Trondheim, Bergen, and Oslo Municipality currently all have small-scale "arts on prescription" projects including populations such as elderly and mental health (Jensen et al. 2016a).

The Danish Ministry of Culture does not mention arts and health in its 2013–2016 strategy. Mental health and interdisciplinary work across sections are however mentioned in the strategy for health. There is no documentation on the effects of culture, arts, and health projects in Denmark. However, there has been research in related fields in Denmark such as research on music and public health. In a national survey in 2014, 14,000 Danish citizens replied to questions on the use of music in everyday life in relation to the individual perception of health (see Chap. 2 by Ekholm and Bonde). Building on experiences from Skåne in Sweden and after much debate and public state hearings in 2016, the Danish government funded a 3-year pilot project to evaluate the effect of participation in cultural activities for citizens experiencing anxiety, depression, and stress. This project called "arts on prescription" has made several municipalities join forces with local cultural institutions to try out and document arts and health projects in Denmark. Research on the effects and properties of arts and health projects in Denmark mainly has its focus on music therapy and music situated within the music therapy environment at Aalborg University, Aalborg University Hospital, and Center for Music in the Brain at Aarhus University. A recent qualitative interview study on the experience of participating in a structured museum visit for people with mental health issues reveals beneficial health-promoting outcomes such as increased well-being and quality of life. The study also highlights that culture providers need training to be able to meet the needs of vulnerable populations and to ensure maximum health-promoting value in cultural activities (Jensen et al. 2016b).

Recently, a much needed research center specifically for arts, culture, and health in Denmark was formed by Aalborg University in partnership with Aalborg University Hospital, Aalborg Municipality, and the North Denmark Region in a cross-sector collaboration (Jensen et al. 2016a).

Culture on Prescription and the Use of Music

As mentioned above, the Danish government has funded a 3-year project called "arts on prescription" for long-term sick leave citizens with mild to moderate depression, anxiety, and stress. The purpose of this project is the following:

- To test how cultural and art activities can be included in public services within long-term sick people with mild to moderate depression, anxiety, and stress
- Based on experience from the project, to develop a model for how cultural activities can be included in the rehabilitation of citizens in a Danish context

Arts on prescription has set out to be holistic and focus on all aspects of the individual including feelings, impressions, and reflections of participants with the emphasis on forming a community with others in cultural activities. The dynamics between the participants are meant to give each of them the opportunity to be part of a larger community and let illnesses fade to the background. By reflecting and talking about the cultural experiences, participants can remobilize mental resources and thereby maybe cope with their illness and challenges in a better way.

The activities all have cultural content within artistic areas of music, arts, building, landscape architecture, and literature. Like in Skåne, the participants are offered various activities in groups of about ten participants three times a week for a period of 10 weeks. The course has a variety of different cultural activities, where participants can take an active part and reflect on their experiences. Unfortunately, less consideration is given to the documentation and evaluation of the 3-year project. The pilot project includes a prepared questionnaire to collect data on the participants' experience of the activities. However, a description of the local success criteria is also required as part of the evaluation looking at what seemed relevant for the organizational and substantial aspects of the effort including partnerships developed between administrations and cultural institutions. Information about data collection and specific outcome measures has been provided late in the project, which has made it difficult for municipalities to plan for the data collection as a natural part of the activities (Sundhedsstyrelsen 2016).

"Culture Vitamins": An Aalborg Model

Aalborg Municipality together with many other municipalities has made a tremendous effort to meet the demands of the Danish Government, and they have managed to make agreements with a number of cultural institutions and activities:

- The Rhythmic School in Nordkraft – focusing on singing together
- Music as Coping Strategy – customized playlists for mental breaks, sleep support, and the conscious use of music
- Libraries – guided literature reading in groups
- Artwork – experiencing art and participating in workshop at KUNSTEN Art Museum
- Theater – actors providing insight into acting and body language plus theater events
- City Archives – genealogy and historical knowledge about the home city
- The Aalborg Symphony Orchestra – participation in rehearsals and introductions to concerts

Music plays a recurring role in the course for the group – both as active listeners and as active participants. The course lasts 10 weeks, with three seminars per week consisting of a variety of cultural activities.

The Rhythmic School in Nordkraft – ten seminars once a week to ensure that participants experience predictability and recurring characters in the process. The course is about singing with others and experiencing the joy of singing and vocal expression as well as being part of a community. Singing prerequisites are not required.

Music as Coping Strategy – three seminars in total placed in the opening, interim, and final seminars. The Music Therapy Clinic at Aalborg University Hospital has been chosen to perform this task due to their extensive experience in working with music and its effect on people with mental health problems.

Libraries in Aalborg – eight seminars with approximately one session per week for participants to experience predictability and security and get familiar with visiting the library.

KUNSTEN – two seminars. KUNSTEN is Aalborg's internationally renowned art museum, which also organizes workshops. The process of visiting KUNSTEN consists of a seminar where participants experience the museum's collections and are meant to discuss their experience and another seminar in which participants are in a 3-h workshop working with art in the museum's workshop.

Aalborg Theatre – two seminars. Participants are introduced to theater performances and the theater profession before, during, and after a performance. They get insight into acting, how the body communicates, and how to use this in everyday communication.

City Archives – about three seminars. City Archives provides tools to understanding the local surroundings of citizens in the past, the present, and the future.

Aalborg Symphony Orchestra – one to two seminars in which participants after preparation at the library take part in a few concert rehearsals and are introduced to the music, the instruments, their role, and the history.

Aalborg Municipality provides a main coordinator participating in all activities together with the groups, ensuring trust building in the group and constant evaluation and adjustments throughout the course. The Health and Culture Department at Aalborg Municipality has a lot of experience in combining the two fields as one of their basic strategies is about creating synergy between health and culture. With Culture Vitamins, the focus is to offer people with mental challenges a well-documented service that can establish communities and facilitate ideas of how to create moments of peace and opportunities to put their illness to the background and themselves and their self-esteem to the foreground (Aalborg Kommune 2016).

A unique part of the program established in Aalborg is the inclusion of music listening at home initiated and evaluated in three seminars. The title of the course is "Music as Coping Strategy," and this course is undertaken by music therapists. The following section gives an introduction to the experiences from a psychiatric unit prior to the inclusion of this particular course in Culture Vitamins.

Music Listening: Prior Experiences

When the initiative to bring the concept of culture on prescription to Denmark was approved politically, a group of music therapists in Aalborg saw the opportunity for music listening to be included in the project and contacted the regional organizers in northern Denmark. At the Psychiatry Unit, Aalborg University Hospital, several sound and music intervention projects have been initiated since 2007 (Bonde 2009; Lund, 2011; Lund & Dammeyer Fønsbo, 2011; Schou, 2007; Schou et al. 2011). Music listening with the use of sound pillows as well as newly developed music equipment is commonly used in treatment by nurses and caregivers in many inpatient units in psychiatry as well as in the postanesthesia care unit in somatic care. This practice has been established by music therapists in the psychiatric field, as well as in the somatic field by a medical doctor with a special interest in music and sound milieu, Dr. Per Thorgaard (Thorgaard et al. 2005). Music on prescription or music medicine is therefore already an existing phenomenon in clinical practice in psychiatry. Citizens who have been admitted to the psychiatric hospital in Aalborg with problems related to stress, anxiety, and depression have had the offer to listen to music for relaxation and for sleep support during their hospital stay (Hannibal et al. 2013; Lund & Bertelsen, 2015; 2016; Thorgaard et al. 2005).

In a recent initiative providing music listening to patients in psychiatric intensive care, the music therapists at Aalborg University Hospital have developed a new application (app) for iPad called *The Music Star*. The Music Star is a user interface in the form of a star, making music choice simple and inviting. The Music Star comes with a number of playlists specially designed by music therapists based on the playlist's calm and relaxing features (Lund et al. 2016).

For the group of citizens included in "Culture Vitamins," the selection of calm music in the playlists is developed for a hospital setting. In general, the participants are sensitive to stimuli, and some would seem to be suffering from stimulus overload in their daily life, causing them to be in a more or less constant state of alert, hence having difficulties achieving peace and rest. The music in the playlists is mainly chosen for its minimal or limited degree of musical challenges and with the aim of lowering arousal, thus providing possible relaxation and focused attention for the listener.

Music as Coping Strategy

In the fall of 2016, the first groups started the 10-week course. As mentioned earlier, the program includes a course called "Music as Coping Strategy" led by two music therapists. Significant for the "Music as Coping Strategy" course is the fact that music listening is an activity the participant is instructed to perform individually at home. Other cultural activities in "Culture Vitamins" are characterized by a social dimension, requiring the individual to leave home and to engage in group activities with other people. For the group of vulnerable citizens participating in "Culture Vitamins," it is important to counteract isolation by offering cultural activities in social contexts. On the other hand, it is equally important to respect the time and pace of the individual and to not push the recovery process. An option is to then bring the cultural activities to the individual at home and to offer cultural activities in the safe surroundings of their home. When being on sick leave, the citizen spends most of his or her time at home. The motivation to engage in music listening at home does not require motivation and effort to get out and socialize. Moreover, music listening is often perceived as a positive everyday activity (MacDonald et al. 2012). This can support the participant's motivation to use music as a strategy in their daily life.

The Seminars

The course aims to empower the individual to make constructive choices using music in the recovery process according to three guidelines:

1. Taking a musical break (offering distraction from thoughts and worries)
2. Using music according to individual needs (by "self-prescription")
3. Using music as sleep support (playing music before going to bed or when not being able to fall sleep)

The seminars offer discussion and dialogue on how music is used by the participants, guidelines on how to use music in a health perspective, and music listening exercises combined with relaxation techniques, visualization, and exercises from cognitive therapy. The first seminars include technical instructions on the use of the equipment, which the participant can bring home during the full length of the Cultural Vitamins course. The music therapists involved in the course instruct and guide the participants in using music as a strategy, but do not act as therapists. In the following seminars, the participants are encouraged to share their personal experiences of music listening with the group both with a learning aim and a social aim. Sharing is helpful as it illustrates individual and creative ways of using music to improve and promote health.

Sound Equipment

Throughout the entire length of the course, each participant has an Apple iPad Mini 2 with the *Music Star* app installed and a BOSE SoundLink Bluetooth Speaker III. This speaker is carefully selected among others for its rich and warm sound even at low volume, its long battery life, and easy transportation, enabling the participants to use it in different rooms and even use it away from home. In addition, the chosen speaker is very easy to use, the buttons are large, and its functions are simple to use. The *Music Star* holds approximately 10 h of music divided into 14 playlists. The length of a playlist is between 30 and 60 min, and its design makes it easy, quick, and intuitive to use. Each playlist is represented by a colored triangle in shades of blue, green, red, and gray, altogether forming the shape of a star, and is activated by finger touch. A taxonomy of music stimuli (Wärja and Bonde 2014) has been incorporated into the design, using the first third of the taxonomy. The music represented here is characterized by low complexity in composition, instrumentation, tempo, rhythm, harmonies, melodies, etc., all providing predictable music and providing low arousal for the listener. The *Music Star* is providing an increase of musical stimuli, going step by step clockwise from low stimuli in the lightest blue triangles to more stimuli in the red triangles. The design facilitates the intuitive and simple use of the *Music Star* and at the same time aids the listeners to remember their preferred music and playlists (Fig 13.1).

Fig. 13.1 The *Music Star* app. The app provides an intuitive user interface to the playlists, arranged in a color system where the level of stimulus (from very low to very moderate) is allocated to the colors (from bright blue to red)

Moreover, the app contains an advanced log file function, enabling researchers to do a statistical analysis of the specific use of the *Music Star* by each participant and the group as a whole. These results will be available after the 3-year period of the course when approximately 300 participants will have concluded their courses. For a more elaborate presentation of the *Music Star*, see the article in *SoundEffects* (Lund et al. 2016).

Music Therapist as Culture Guide

When facilitating the groups, the focus is to use music as a coping strategy, and the role of the music therapist is to be course leaders or "culture guides" (this term is used in the written material for the participants). The rationale for the concept of the group work is different from that of group music therapy considering the role of the music therapist and of the client/participant.

Firstly, the group meetings are not named music therapy. Secondly, the music therapists who are employed as culture guides are not engaging as therapists, and supervision is not included in the concept. The music therapists are engaging to help participants use music as a tool for relaxation, a distraction from stressful/negative thoughts, and sleep support. Looking at the role of the music therapist in this context, it is relevant to look at the setting and description of the course.

The music therapists educated at Aalborg University often have an eclectic approach based primarily on a psychodynamic theory and method (Bonde et al. 2014). This course has a behavioral aspect using music for emotional regulation by the individual. The relational focus is in the background, and the behavioral aim using music to regulate mood and emotion is in the foreground. This approach is influenced by the concept of "music medicine" as described by Bonde (2014). The original title of the course "Learn How to Listen to Music" was recently changed to "Music as a Coping Strategy." The introduction text for the participants is short and says:

> Most people agree that music has an influence on mood. The music preferences and listening experiences are individual. Music listening can be helpful for having a mental break, for relaxation and for sleep support. Music therapists guide the participants to use The Music Star – a new tool to make music choice. Course duration 2 hours. (Translation by authors)

When working as a "culture guide," it is important to focus on the present reality and on issues related to the use of music for the well-being of the individual. It is of less importance to understand the individual history of the participants. Information and details concerning key problems, understanding of self and the current situation, pathology, diagnosis, hospital admission, and family relations that arise in the group are not subject to further investigation. This means that the therapist/guide will not question or initiate discussion on these topics but will engage in empathic listening to the themes when brought up by the participants. The music therapists will direct the attention to the details that are helpful to understand present issues concerning music listening practices. They will not ask questions that motivate to explore personal material not related to music listening. Questions that may be posed concerning the past are: "How did you listen to music before?" "What is different in the way you listen to music now?" "Did you listen to music before in a way that you would like to do again – or that you now want to avoid?"

Culture guides with a background in music therapy with a relational focus may act and speak in a way that promotes reflections and dialogue in a psychotherapeutic understanding. This may happen in spite of the culture guides' effort to withdraw from the use of therapeutic language. Issues often brought up by the participants have been (1) challenges concerning the engagement in music listening, (2) steps toward getting better, and (3) functional level. Questions concerning music listening habits can easily lead to existential themes like "Who was I before? Who am I now? Do I want to be as I was before?" This may lead to further dialogue on change and personal process among the participants. Inevitably, personal development is a key issue for the participants, and the talk during the group meetings does have similarities to the dialogues in psychiatric outpatient music therapy groups.

Some participants look at their past abilities with sorrow wishing that things will return to the way they were before their illness. The past reminds them of better times. Some participants see the need for change and look back at an unhealthy lifestyle with sorrow. The past reminds them of things that went wrong. When wishing to return to the past, the participants tend to prefer to listen to music they know as it brings them good memories. When the past is painful, participants have strong wishes to search for new music since they are searching for new healthier ways of living. These observations are not drawn on theory but on the experience of being a culture guide.

Being Part of a Group

The rationale behind guiding a group is the wish to provide a secure learning forum where the participants can share their personal experiences with music listening and be inspired by others in the group. Sharing in a group can be challenging for some at first but can also be very supportive and trust building. Some of the early groups have upon ending the course made arrangements to stay in touch and meet for future social and cultural events.

Participant Testimonials

(a) I use The Music Star with my children (4 and 8 years old) in the afternoon when I need a break. First, it lasted 5 minutes only, but now we have a break for up to 20 minutes.
(b) I am very pleased that The Music Star is included in the program because it gives some surprising experiences and even epiphanies.
(c) The music I use on The Music Star is a kind of music I never listened to before. You have to make friends with this kind of music. It takes an effort to get used to the fact that the music which is good for me now is a kind of music that I didn't listen to before. (Translation by the authors)

Another example of a personal creative and helpful way of using music listening as a coping strategy is a woman who describes being paralyzed by stress in the morning hours when thinking of all the insurmountable things she is expected to do during the day. She decides to listen to a particular playlist in the kitchen in the early morning hours telling herself: "I am allowing myself not to start the day yet." By creating that space with the music, combined with the instruction that the day and the worries and activities haven't started yet, she gives herself a moment of freedom, a needed break from bodily and mental stress. This musical morning pause makes the start of the day less stressful and gives her confidence that she will be able to manage the challenges during the day.

Other participants report that the use of specific music enables them to raise or lower their arousal, to fit or induce the energy level needed to manage daily tasks such as cleaning and hoovering or simply having a mental break or a pause. Even though carefully chosen music and sound equipment are offered, it is not always beneficial for everyone, and some participants found the music to have too much stimuli, and others to have too little. Some didn't like portions of the selected music, and some didn't want music at all.

One participant states, "The music is like soothing warm oil on my body," while another finds the music to be too quiet and does not use it much. Her primary strategy for stress relief instead is to knit or to do crochet, and she does not seem to have the need for more strategies. The vast majority of the first four cohorts found the music to be very helpful. They adapt the techniques taught during the seminars and even found new personal ways of using the music. Many participants also point out that they regarded music differently before and after the course. Today they are much more aware of the choices they make when listening to music in general.

One of the participants was extremely sensitive to sound and noise and therefore constantly in a state of alert, followed by severe fatigue later during the day. A suggestion by the music therapists to use earplugs or headphones in order to shield herself and control her level of stress and anxiety becomes a turning point and a great relief for her. Sometimes very simple measures can be of great help.

Although the effects of music listening can be generalized, it is important to remember that it is also a very personal experience, in which mental state along with personal memories and feelings attached to the specific piece of music plays a role. Music suggested by the music therapist can therefore sometimes be overruled by personal preferences that might seem to be inconsistent or even contraindicated.

Discussion and Conclusion

There is currently much engagement and excitement around the new political focus on developing arts, culture, and health as an active field in Denmark. Drawing on the knowledge and experience from our closest neighbors is highly relevant, but we still have a lot to learn as each country has a different healthcare system and a different view on arts and culture and health. It seems relevant to discuss how we can best implement arts and health in Denmark and what role music and the music therapy profession can or should play in this process. What are the challenges and potentials and how can we unitedly secure a clear sustainable national strategy?

The project in Skåne is very impressive and inspirational, and Denmark can be grateful to be able to draw on these Swedish experiences. However, as the Danish State in their project chose to focus on a different and more vulnerable population such as long-term sick leave citizens with mild to moderate depression, anxiety, and stress, one might have considered adjusting the format and amount of seminars concurrently. Some of the citizens in Aalborg understandably could not cope with seminars three times a week for 10 weeks. Quantity and outcome quality is a challenge and cost benefits need to be of greater concern in future research studies. In Skåne the experimental group was compared with a historical control group looking into records of sick leave and general healthcare costs. However, the control group did not answer questions on the quality of life and possible improved social/communication skills. There was also no active control looking into what specific role the art experience had compared to sharing in a group nor a focus on groups receiving different "doses." Without this information, it is difficult to know what kind of activity works better for a specific need of the challenged citizen and to know how much is needed. A control group cannot easily be established within municipalities, as this is not their familiar way to evaluate projects. Ethics become a challenge as expensive research goals rarely are prioritized over basic needs of citizens which is why both state and research institutions have to support and secure strict scientific evaluations of arts and health projects including both effect studies and qualitative studies on how and why it works.

The title of the course in Culture Vitamins is "Music as Coping Strategy." It is important to draw attention to the term "coping strategy." Using music listening in relation to coping with mental problems implies that music can be used as a tool for coping. The use of music as a tool in this situation requires a conscious choice by the individual to consider and actually use music as a tool to obtain a particular effect. The calming and stress-reducing effect of the selected music made available through the *Music Star* can promote health. The calm and slow music reduces the heart rate and has the potential to influence physical and psychological well-being. Reflecting on this, it seems "Music as Coping Strategy" as the only course in the program gives the participant a tool to use consciously in health promotion. In addition, this course is not solely based on social activities but also includes individual experiences and empowerment. It will be an interesting issue for further investigation how this will be perceived by the participants in future courses.

Music therapists are trained professionals who are skilled in using music to improve and promote health in clinical and nonclinical settings. Professional musicians with a background from a conservatory/music academy or the university may have qualifications to act as cultural guides facilitating music listening groups and encouraging participants to listen to music as a positive and constructive activity in a nonclinical setting. Nurses and caregivers in the intensive care units at Aalborg University Hospital are also helping patients to select and play calm music to reduce anxiety when admitted to hospital. Clearly, it seems appropriate and necessary that professionals other than music therapists contribute to the promotion of music in healthcare. However, previous pilot studies from a psychiatric hospital setting point out that instruction of caregivers by the music therapist and the professional expertise in selecting music is an essential part of the music interventions (Hannibal et al. 2013). These results are not based on music listening practices in nonclinical settings, and therefore they may not be transferable. Still, music therapist competences seem to be important to consider. The music therapist is conscious about the many factors influencing the music listening experience including psychological, physical, and social factors. It is clear that the field of music therapy can contribute with valuable insights into how artists or culture guides can facilitate good effective arts experiences for challenged citizens. Music therapy researchers can also offer insight into how to examine, document, and understand the effect of actively listening to music as part of a recovery and empowerment process.

References

Aalborg Kommune. (2016). *Kulturvitaminer. Information om kurset til rådgivere*. Aalborg Kommune.

Aasgaard, T. (Ed.). (2006). *Musikk og helse*. Oslo: Cappelen Damm.

All-Party Parliamentary Group on Arts, Health and Wellbeing. (2017). Creative health: The arts for health and wellbeing. Online Publication. http://www.artshealthandwellbeing.org.uk/appg-inquiry/Publications/Creative_Health_Inquiry_Report_2017.pdf. Accessed 06 Nov 2017.

Batt-Rawden, K., & Tellnes, G. (2012). Social factors of sickness absence and ways of coping. A qualitative study of men and women with mental and musculoskeletal diagnoses. *International Journal of Mental Health Promotion, 14*(2), 83–95.

Bonde, L. O. (2009). Lydpuder med musik til psykiatriske patienter – alternativ PN-medicin? *Psykiatrisk Sygepleje, 17*(1), 14–16.
Bonde, L. O. (ed) (2014). *Musikterapi – teori – uddannelse – forskning – praksis [Music Therapy – Theory – Education – Research – Practice]*. Aarhus: Klim.
Fancourt, D. (2017). *Arts in health: Designing and researching interventions*. London: Oxford University Press.
Hamilton, K., Buchanan-Hughes, A., Lim, S., & Eddoweset, L. (2015). *Evidence Dossier The Value of Arts on Prescription Programmes for the Mental Health and Wellbeing of Individuals and Communities*. Cambridge: UK.
Hannibal, N., Lund, H. N., & Bonde, L. O. (2013). Musiklyttepuder, lyd-bøjler og spillelister i behandlingen af psykiatriske patienter. *Musikterapi i psykiatrien Online, 8*(2), 4–17. https://doi.org/10.5278/ojs.mipo.v8i2.553.
Jensen, A., & Wille, G. G. (2016). A Danish perspective of culture and health – Towards an interdisciplinary approach. *Journal of Applied Arts & Health, 6*(2), 129–138.
Jensen, A., Stickley, T., & Edgley, A. (2016a). The perspective of people who use mental health services engaging with arts and cultural activities. *Mental Health and Social Inclusion, 20*(3), 180–186.
Jensen, A., Stickley, T., Torrissen, W., & Stigmar, K. (2016b). Arts on prescription in Scandinavia: A review of current practice and future possibilities. *Perspectives in Public Health, 137*(5), 268–274. https://doi.org/10.1177/1757913916676853.
Lund, H. N. (2011). Musikterapi med musiklyttepude på lukket akut modtageafsnit. *Tidsskriftet Dansk Musikterapi, 8*(1), 28–29.
Lund, H. N. (2016). *Pilot project: Sound pillow treatment to improve sleep quality for patients with depression and sleeping problems*. Proceedings of 24th European Congress of Psychiatry, Madrid. https://doi.org/10.1016/j.eurpsy.2016.01.026.
Lund, H. N., & Bertelsen, L. R. (2015). *Music listening as anxiety management in intensive psychiatry*. Proceedings of the 9th Congress on Violence in Clinical Psychiatry. Dwingeloo: Kavanagh Publisher.
Lund, H. N., & Bertelsen, L. R. (2016). Musikstjernen – en ny stjerne i psykiatrien. *Tidsskriftet Dansk Musikterapi, 13*(1), 3–10.
Lund, H. N., & Dammeyer Fønsbo, C. (2011). *Musiklyttegrupper – en empirisk undersøgelse af anvendte metoder i psykiatrien. Musikterapi i psykiatrien, Årsskrift 6(1)* (pp. 86–101). Aalborg: Musikterapiklinikken. https://doi.org/10.5278/ojs.mipo.v6i1.103.
Lund, H. N., Bertelsen, L. R., Bonde, L. O. (2016). Sound and music interventions in psychiatry at Aalborg University Hospital. *Sound effects*, http://www.soundeffects.dk/article/view/24912/21827. Accessed 30 Nov 2017.
MacDonald, R., Kreutz, G., & Mitchell, L. (2012). *Music, Health, and Wellbeing*. Oxford: Oxford University Press.
Schou, K. (2007). Musikmedicin og musikterapi i medicin. *Psyke & Logos, 28*(1), 525–547.
Schou, K., Pedersen, I. N., & Bonde, L. O. (2011). Musiklytning til patienter i skærmning. Pilotundersøgelse på Musikterapiklinikken. Musikterapi i psykiatrien. *Årsskrift, 6*(1), 56–67. https://doi.org/10.5278/ojs.mipo.v6i1.101.
Stigmar, K., Åström, M., Sarbast, S., & Petersson, I.F. (2016). Kultur på recept 2.0. www.skane.se/siteassets/kultur/dokument/utvardering-kultur-pa-recept-april-2016.pdf. Accessed 23rd Jan 2017.
Sundhedsstyrelsen (2016). Udmøntning af satspuljen. Kultur på recept. https://www.sst.dk/da/puljer-og-projekter/2016/~/media/88ECB7DFA7BC47DAB89EEB8FC8BEF518.ashx. Accessed 24 Jan 2017.
Thorgaard, P., Ertmann, E., Hansen, V., Nørregaard, A., Hansen, V., & Spanggaard, L. (2005). Designed sound and music environment in postanaesthesia care units—a multicentre study of patients and staff. *Intensive and Critical Care Nursing, 21* (4), 220–225.
Wärja, M., & Bonde, L. O. (2014). Music as co-therapist: Towards a taxonomy of music in therapeutic music and imagery work. *Music & Medicine, 6*(2), 16–27.

Chapter 14
The Fellowship of Health Musicking: A Model to Promote Health and Well-Being

Kari Bjerke Batt-Rawden

Introduction

Public Health Challenges and Their Link to the Benefits of Cultural Participation

In recent years a comprehensive public health perspective has been based on a broad and inclusive conceptual framework of positive health and health development. There are good reasons to consider music – and cultural practices more widely – in connection to health. Music is often an unrecognized resource of social agency, a medium of social order and practice (DeNora 1999, 2000). Even though music has been considered as a contributor to our health and well-being (Ansdell 2004, 2015; Batt-Rawden et al. 2005; Batt-Rawden 2007, 2010; Batt-Rawden and DeNora 2007; Bonde 2011; Bonde et al. 2013, DeNora 2000, 2014; Ruud 2001, 2002, 2005; Stige 2005, 2012; Theorell 2014), music is basically absent from any political, national or local community discussions about health and quality of life; hence its powers are still underrated.

In the Nordic countries, mental health problems combined with somatic diseases are public health challenges along with an increase in non-communicable diseases[1] that need to be met with health-promoting initiatives and public health strategies (Batt-Rawden and Tellnes 2011, 2012; Tellnes 2003, 2005; Tellnes et al. 2017). The Lancet Commission: Culture and Health (Napier et al. 2011) suggests that avenues to understand and nourish well-being should become the highest healthcare priority and a focus for researchers. The results support hypotheses on the beneficial effect

[1] Also known as chronic diseases, tend to be of long duration and are the result of a combination of genetic, physiological, environmental and behaviours factors (http://www.who.int/mediacentre/factsheets/fs355/en/).

K. B. Batt-Rawden (✉)
Department of Health Science, NTNU, Gjøvik, Norway
e-mail: kari.batt-rawden@ntnu.no

L. O. Bonde, T. Theorell (eds.), *Music and Public Health*,
https://doi.org/10.1007/978-3-319-76240-1_14

of cultural activities in health promotion and healthcare (Konlaan et al. 2000; Tellnes et al. 2017). There are also examples of the effects of engaging with cultural activities 'levelling off' health inequalities (Napier et al. 2011). The Norwegian HUNT study found significant positive associations between participation in cultural activities and perceived health, improved life satisfaction and reduced anxiety and depression (Cuypers et al. 2013; Hansen et al. 2015) (see Chap. 3). If health in this context can be defined as having the energy to cope with the tasks and challenges of everyday life (Tellnes 2003), in spite of symptoms and diseases, recent research shows how cultural practices may promote healthy behaviour through salutogenic (health causing) factors (Tellnes 2005).

Music's Power to Promote Health and Well-Being: A Salutogenic Approach

The theory of salutogenesis (Antonovsky 1996; Lindstrøm and Eriksson 2015; Mittelmark et al. 2016) is the overall paradigm in The Fellowship of Health Musicking Model. Salutogenesis identifies perceived individual and collective general resistance resources (GRRs). GRRs are physical and biochemical resources, knowledge, identity, social support, inner feelings, values, coping strategies, participation in nature and culture activities and different material goods – resources that may promote an effective management of tension in demanding situations. The focus in salutogenesis is how to enter into a good circle, a positive feedback loop (Langeland et al. 2007; Langeland and Wahl 2009). According to Lindstrøm and Eriksson (2015), it is obvious that a salutogenic mindset can provide an effective approach to health education, and they emphasize that it is difficult to understand why this thinking and acting is not used more than it is today.

Music's beneficial effect seem to be created in and through health musicking (Stige 2012, see Chap. 8 by Stige) which also can be understood as any use of music experiences to regulate emotional or relational states or to promote well-being, be it therapeutic or not, professionally assisted or self-made (Bonde 2011, p. 140). Music's real effect seems to be its specific ability to provide mechanisms and affordances for achieving those healing effects (DeNora 1999, 2000). As music is a ubiquitous medium and very much embedded in everyday life, once people are encouraged to learn how to use music's power as a health resource, it may be the most pragmatic, cost-effective and sustainable means that can be utilized in public health (Brunges and Avigne 2003; Gagner-Tjellesen et al. 2001; Maranto 1993). In line with previous research, music's role as a resource for configuring emotional and embodied agency is *not* one that can be predetermined (Batt-Rawden 2007; Batt-Rawden et al. 2007; DeNora 2000; Schäfer et al. 2013). Music is personal, albeit culturally patterned, and cannot be prescribed in a manner akin to pharmacology (Sloboda 2005). That is, the individual variability in responses to music makes a generalized application of music as a treatment 'like a pill' very difficult. This is why in most cases *self-selected* music exerts the greatest positive effects on listeners and works as a *catalyst* (Chanda and Levitin 2013; Batt-Rawden 2010; Elvers 2016). It seems to be in the transcendental musical moment one actively performs self-health; hence 'high points' or 'peak experiences' or even ordinary musical experience may contribute to healing, recovery and feelings of self-efficacy (Batt-Rawden 2007; Fancourt et al. 2014; Gabrielsson 2011; Juslin and Sloboda 2010).

This chapter discusses a new approach to promoting health and quality of life in local contexts by learning participants to use music as a 'technology' of health and self-care through the steps and actions of The Fellowship of Health Musicking Model. *The main purpose* of this chapter is to increase knowledge as to firstly how musical activities can promote mental and somatic health and, secondly, how The Fellowship of Musicking Model may be used as a health-promoting initiative. The *model* builds on a novel musical health promotion procedure developed by the author in 2007 as part of a PhD project. The aims of the study were, firstly, to explore the role and significance of music in the life of men and women with long-term illnesses in or through different life phases, situations, events, issues and contexts and, secondly, to increase knowledge on how participants, through exposure to and exchange of new musical materials and practices, may learn to use music as a 'technology of self' in relation to health and healing. The longitudinal study[2] involved 9 men and 13 women, aged between 35 and 65, and was a pragmatic synthesis of elements of ethnography, grounded theory and action research. Eight in-depth interviews were conducted with each participant, and open narratives were elicited from each of them, using a topic guide, two single CDs and four double compilations. The study drew upon participatory action research to the extent that it sought to encourage participants to bring to the level of conscious activity their various uses of music for the care of self and to instigate such practice and reflection.

Description of How the Model Works as a Practical Method

The Fellowship of Musicking: A Model to Promote Health and Well-Being

The Fellowship of Musicking Model can hopefully inspire and motivate facilitators[3] to use, adopt, adapt or develop this model as a tool to promote health for actors in local communities and patients in hospitals and health centres, institutions or rehabilitation centres. The model runs over a period of six sessions lasting for 2 h once a month and includes specific lessons or homework during the five intervals of

[2] This chapter includes both published and unpublished material from the PhD thesis (Batt-Rawden 2007).

[3] Facilitators: general practitioners, nurses, public health officers, music therapists, social workers or equivalent health professionals.

these sessions.[4] It is designed for small groups or individual sessions depending on context, issues or health challenges of the actors. Music sharing may also turn out to constitute a context that creates a space for individual-based aesthetic exchange.

The Informal Learning of Health Musicking

Before taking the reader through the six sessions and specific steps in more detail, a vital part of the model contains five intervals between the sessions. That is, the participants' homework or lessons are part of the informal learning of health musicking (Batt-Rawden and DeNora 2007; Green 2002; Lindstrøm and Eriksson 2015; Savage 2005) and are ways to instigate and motivate empowering rituals through the act of listening and the act of choosing self-selected music. As previous findings highlight, the emotional, spiritual, participatory and bodily engagement or attachment with the music seem to be important prior to agency or any kind of faith healing. That is, participants identify what works and learn how to invoke and empower that musical 'charm' (and themselves in relation to it), and -it works. In a very real sense, quality of life, even at times, 'being ill' or 'being well', takes shape in relation to the resources for self-construction, and one can learn how to access these resources in constructive ways, and this is the learning process (Batt-Rawden and DeNora 2007; Batt-Rawden 2010).

Previous findings do not imply that music 'causes' the participants to behave in certain ways; rather, they show how music's power acts as an aesthetic technology (DeNora 2000, 2014), one that can 'transport', 'lift' or 'transfer' an individual from one state of mind or emotional position or height to another [better]'dimension'. This transportation seems to be achieved through a deep, focused, concentrated musical experience, here to be obtained through the act of listening and choosing procedure as part of their homework in everyday life settings, i.e. a kind of 'deep' musicking[5] or 'deep' self-musicking[6] that demands focused attention over time, a form of musical mindfulness. These concepts are close to how Becker (2004) describes *deep listening* or Green's (2002) *purposive listening* as skills that are life-enhancing and enriching and which often bring deep rewards and may reduce anxiety and pain (Theorell 2014). These moments of musical mindfulness and high points are of great importance for participants' sense of well-being and recovery and sense of being restored into wholeness and could be related to a relational problem, an illness, a grief, a sadness, a depression or anxiety or a 'need' to celebrate health revisited or regained.

---

[4] The participants' narratives used as illustrations here partook in individual sessions (Batt-Rawden 2007).

[5] The two concepts developed here are based on Small's concept of 'musicking', Becker's 'deep listening' (Becker 2004) and Green's (2002) 'purposive listening'; 'deep' musicking: deep, focused or concentrated musical activity over time.

[6] 'Deep' self-musicking: deep, focused or concentrated singing or playing for self or others over time.

These musical practices seem to be tacit, revealed only when things are wrong. In consequence, the lay knowledge of self-healing as beneficial for care of self is learned and embodied in the 'doing' of these moments of musical mindfulness to be replicated whenever a need occurs. When introduced to musical learning situations, formal or informal, music becomes an arena for the development of self-efficacy.

Planning the First Session

Before the first session with the participants, it is important that the first music compilation (from CDs, Spotify or similar) is musically chosen by the facilitator as to start a sympathetic entry into the participants' social-musical worlds (Batt-Rawden and Aasgaard 2006). Moreover, to be able to elicit participants' life stories and stories of being well and being ill by instigating narratives about music's role in everyday life, a music selection themed *Keepsakes and Memories* is compiled by the facilitator at the onset to bring at the first session. The music selection ought to consist of a selection of main genres/styles to secure a variety that might reflect general musical taste according to age, gender, identity, social background, status group, musical practices, tastes and social location. In general, these genres/styles might be folk/country, classical, popular music and jazz/blues, and a good idea is to choose one or two pieces of music from each category, representing different periods in time.

The First Session: Listening to Keepsakes and Memories: Musical Dialogues

The musical pieces from the first theme are included as potential triggers for memory work and thus as potential prompts for discussions and musical dialogues. The first session include questions about their listening practices and are related to their musical tastes, practices, habits and preferences, as to start the process. At the end of first session, the facilitator will give the participants' their first homework related to the second predefined theme to instigate the cyclical process. The task is to listen and to choose two pieces of music that is or has been of particularly significance to him or her.

Homework at the Intervals: The Act of Listening to and Choosing Musical Material

Following the order of the sessions and the task of homework during the intervals, this work relates to participants' *act of listening and choosing musical material related to the following five predefined themes*: 'Music and its significance for me and why', 'My Mood', 'Feeling at my Best', 'Musical Mindfulness' and 'All-Time Best'.

Part 1: The act of listening: Questions to be asked at the session two to six are: Have you listened to [your] music? What have you listened to? Why? How did you begin to listen? Where (room/place/space)? Why? Did you look specifically for something or play tracks randomly, or did you listen straight through? Did you repeat a track? If you did like some (or one), what did you like? Why? What did you dislike? Why? Open narratives – any issues or thoughts you have in mind that may relate your past or present life situation? How?

Part 2: The act of choosing: Questions to be asked at the session two to six: How did you come to choose what you chose? What is/was it about the piece of music you have chosen? Would you be able to describe your mood when you chose which piece of music to share and listen to? Why? Your current life situation? If so, how and Why? Is it possible to connect your current life situation with your choice of music? If so, why? How? Are there any events since last time we met that might relate to the chosen piece of music? What? Why? How?

Both the *listening and choosing procedures* is in line with Charmaz' description (1999:378) of how significant events may reveal images of present or possible self and evoke feelings. The 'themed' music will construct and create the participants' reflections of relevant issues, events and situations, allowing significant associations to emerge at the sessions. By requiring the participants to choose their own music, one gets closer to their biography and life stories and self-reflexivity. Through their personal chosen pieces of music, the participants may start to draw connections between musical materials and situations of health or illness.

The Second Session and Theme 'Music and Its Significance for Me and Why'

At all sessions from now on, it is important to play and discuss the selected and chosen music from each of the participants' homework related to the theme. By encouraging the participants to 'narrate' and thus learn the skills of telling to self and others how music 'works for me', this theme opens up for narrative opportunities where the participants are asked to engage in talking about music's powers 'for them'. Musical associations seem to be linked to memories of significant relationships and the unique qualities and intensities of these relationships (Butterton 2004). Baumgartner (1992) also discusses of how people often identify specific pieces of music with memories of significant events in their lives.

At the end of the second session, the facilitator asks questions related to the third theme for their next homework. The task is to choose two pieces of music that is or have been related to their current mood.

The Third Session and Theme 'My Mood'

First, play and discuss the selected pieces of music from their homework. A specific choice of music might be 'mood-related' to health and illness in specific contexts and life phases. By asking participants' about their mood, Greasley and Lamont (2006:965) show how music preferences in adulthood described music as a catalyst for achieving mood change, mood enhancement and different psychological states of mind.

At the end of the third session, the facilitator asks questions related to the fourth theme for their next homework, following the same procedure. The task is to choose two pieces of music that are connected to contexts, situations or life phases when they are feeling at their best.

The Fourth Session and Theme 'Feeling at My Best'

At the beginning of the session, play and discuss the selected pieces of music from their homework related to the current theme. Music's ongoing parameters as an 'aural mirror' may make the participants reflect and rethink their situation; thus music's 'effect' could provoke negative past experiences, enhancing feelings of being ill at the present. This notion may relate to where people are in the process of recovery or in the transition from illness to recovery, as part of a salutogenic focus (Antonovsky 1996).

At the end of the fourth session, facilitator asks questions related to the fifth theme for their next homework, following the same procedure. The task is to choose two pieces of music that are related to their musical choices mirroring a musical mindfulness which is part of a focused, concentrated period of deep (self-) musicking.

The Fifth Session and Theme 'Musical Mindfulness and Deep (Self-) Musicking'

Again, play and discuss the selected pieces of music from their homework related to the current theme. The growing musical awareness and consciousness at this point are vital, i.e. how music works, how they have used music previously or how they reflect on their own actions in relation to musical habits and preferences in their past or present everyday life. The participants' reflections may signify 'turning points' and vital moments of discovery towards a healthier self.

Experiences of pleasure motivate deep musical listening and involve skills that are life-enhancing and enriching and ones which often bring deep rewards (Becker 2004).

At the end of the fifth session, the facilitator asks questions related to the sixth and last theme for their next homework, following the same procedure. The task is to choose two pieces of music that are related to their musical choices when they are feeling all-time best.

The Sixth Session and Theme 'All-Time Best'

At the last session, play and discuss the selected pieces of music from their homework related to the current theme. Ask the participants about their listening practices and activities related to this theme and previous themes, and discuss how and why they came to choose what they chose. As to give a sympathetic exit, one might motivate participants for further socio-musical action and networking, thus, as a way to link up informal or formal music-making in the communities.

Previous Results from The Fellowship of Musicking Model

The First Session: Listening to 'Keepsakes and Memories': Musical Dialogues

In this section, results from the author's research following the steps and actions of The Fellowship of Musicking Model are presented (Batt-Rawden 2007). The participants[7] relate their stories to some of the pieces or genres, or they seem to remember them from past musical eras and trends, making comments about their likes and dislikes, associations and memories. In this sense, music is a means of composing a future by creating something new with what is available to us in our context. Ursula narrates:

> I think when I go back in my biography and then play music I have been very fond of, then I feel I send roots all the way down to the sources of my youth…a kind of zest of life, and suddenly I am there again and I feel by opening up those sluices in myself. I get energized and refuelled. [Ursula, age 63. Breast cancer. Post-treatment]

Through this theme one could detect emerging patterns showing how music seemed to be linked to biography, identity or extramusical matters. The participants are encouraged to tell 'analytical stories' that in turn will help to fix music's properties 'over them'. This is to say that music's affordances are constituted through the ways music is framed or prepared for use. To illustrate the process of reflection that may arise, Ursula started to narrate her story while the voice of Eva Cassidy's 'Fields of Gold' was playing in the background:

[7] In this chapter the voices of Edwin, Isabel, William, Ursula, Anthony, Peter, Tara and Alexander are used as illustrations. The different background information is given at the first quotes.

> My life has in a way stopped for 30 years and that influences my relation to my illness and I feel I have a lot of suppressed feelings, and when I start talking about it I also start little by little to work it out, piece by piece and in this way I feel I get better when I listen to music. I think it is through this musical encounter with myself I can untie my unresolved emotions. (*Ursula*)

The Second Session and Theme 'Music and Its Significance for Me and Why'

At this session it is possible informally 'teaching' both themselves and others in the group how to empower music for use. Edwin puts it this way at the second session:

> I have been quite inspired by this project both to seek out new music and to go and buy some new CDs, so I have bought a new CD now by an artist called Frank Britch, which was quite good and some songs of Sibelius which I think are good, but this CD which I would like to select as a piece to go on CD 2, (*showed a CD of Prokofiev*[8]), this is heart music for me. [*Edwin, age 48. Anxiety and depression. Long-term certified sick for a year. Recovered prior to fieldwork*]

When music does 'hit' or play on some 'strings' in their mind and body, they often use music to 'sort things out', a way of ordering social action to support or initiate healing processes. Music seems to be a basic need in Anthony's life when he describes how his health would deteriorate if music were absent. Anthony claims anybody can be good at singing, independent of how well off one is, pointing out how education is irrelevant when it comes to abilities in singing.

> I lost my voice and that was very frustrating, that was one of the worse things, getting a really bad cold and coughing, it wasn't the coughing that mattered, but not being able to sing. It was terrible. I wanted to sing and as soon as I opened my mouth and started singing, I started coughing. I think I would be desperate if it had not cured itself. I suddenly realised how much I missed it, you don't know what you've got till it's gone, just to quote Joni Mitchell. [*Anthony, age 58. Muscular disease. Musically active; singing, playing and composing. Disability pension*]

The Third Session and Theme 'My Mood'

Through 'My Mood' it is possible to see how music works and how participants use music when they are feeling down, for example, constructing healing processes through self-musicking as an important 'meaning-making' ritual:

> What I have noticed is that I am a very bad listener when I feel really down, it does not penetrate, but playing myself [keyboard] does actually give meaning. I remember quite recently I was sitting here in my 'dug-out' and I had just started living alone and was not feeling good at all and I had just bought a keyboard and played a lot on that with different

---

[8] Refer to his choice on the theme 'Music and its Significance for me and why'.

> effects and so on and then I felt that this was quite meaningful to me, because I sort of reconnected myself to my emotions again. (*Edwin*)

At this point several participants started to narrate how their 'mood' relates to 'health of the day', their agency and coping mechanisms. All participants reported using their music preferences to reflect/enhance their current mood or, conversely, to alter/diffuse some aspect of their current mood. Music may be a way of regulating their lives when they are feeling down, tired or weary, and choosing a piece of music may balance their state of mind or mood.

The Fourth Session and Theme 'Feeling at My Best'

Due to the ongoing listening-choosing procedures and exploring in-depth topics that emerge during these sessions, quotes like 'when I am feeling at my best' or 'I go out more' or 'then I play more' or 'I love to listen to this one' are illustrating. Some of the participants had developed musical skills through both formal and informal learning and seemed to understand how and why they experienced musical performances or practices as pleasurable. In relation to how participants seem to link musical choices and uses in situated contexts, this quote also illustrates how choosing a piece of music is embedded in the presentation of self, identity and self-imaging – 'the me in the music' – thus enhancing feelings of well-being:

> This tune enhances my 'feeling good' and it is a happy and simple tune and also the guitar, you know, that is me…I like that kind of guitar playing…high tempo, but even though that tune is simple and the arrangement is quite simple, it has this kind of 'airiness' to it and it has also a sort of relation to the Celtic sound that I like. [*Alexander, age 53. Burn-out syndrome. Long-term certified sick for two years*]

Edwin describes how a piece of music chosen for the theme 'Feeling at my Best'' illustrates his gratitude for meeting a new love and how this music seems to 'speak directly to his emotions', thinking that this simple, expressive and dignified wedding march encompasses his present feelings and 'hits' him in a very special way:

> Well, I have met his woman, you know and I wanted to buy this one (shows me the CD) because I wanted to give you this track, 'Brudesang fra Solør', so it is this mixture with the kind of music I love to listen to and the woman I just met, because it reflects my gratitude for meeting her, music is also so intimate, it is not this kind of distant feeling, it's warm and direct and it just speaks directly to my emotions. I also love Norwegian 'Wedding marches' they can be so wonderful in all their simplistic and simple way, a sort of a dignity and this tune encompasses that, it just hits me. I also like those clarinets. Very expressive. (*Edwin*)

The Fifth Session and Theme 'Musical Mindfulness and Deep (Self-) Musicking'

Several participants sensed how stress, worries and daily troubles were reduced to 'small cares' when they discovered how new musical materials could empower and embody. Having 'adopted' a piece of music as one they could 'try out', this musical 'adoption' might become part of their way to readjust themselves over time. It is through the process of choosing and listening that the patterns of deep musicking and deep self-musicking in relation to their illnesses emerged more clearly:

> I like to continue on that positive wave that occurred to me four years ago, when I found out that I wanted to dedicate my time to music and that is to continue to do the different projects I have in mind or complete little by little what I have already started or it can be just to make new things…[compose] for example I have been thinking for a while what I saw around that lake with the trout where I was this summer and then I want to compose something out of it and I have a lot of those good periods now….[*Peter, age 55. Anxiety and depression. Suffered from a heart failure prior to depression*]

Tara describes how music enriches her, makes her calm and helps her to relax when her life is turbulent or chaotic, by her focus on the importance of being an active listener. She describes her musical ritual consisting of the situational context as one she 'creeps in and out of', almost a trance, consciously making time to do so, often relating to her current mood or present life situation. She claims this might be due to the way she 'works' through her emotions on her own, believing this musical ritual promotes health and quality of life:

> I use music to work things out and as a therapy, to rest, be happy, and feel enriched. I have a tendency to seek out music that makes me rest, makes me calm there and then and I can see a very clear connection between health and music. … so I feel I can crawl into it (the music) and then crawl out again feeling much better, because while I do that, I put aside my life a little bit, you see, and I just don't give a damn about everything else, all that matters 100% then is me in music. I almost enter a trance, a ritual, so I think it actually promotes health to have music as an emotion manager, you know. [*Tara, age 52. Muscular disease; Disability pension*]

Several participants started to make comments on ways they had gained from participating in the project saying, for example, 'I have not thought about music like this before', referring to remembered 'happy times', or 'this has really started a process' or 'I am much more aware how I actually use music now', relating this to why they were moved by certain pieces of music. A majority of the participants admitted that they had not discussed or told anyone else about how they thought or felt about music, life, health or illnesses before. The growing musical awareness and consciousness were a recurrent theme, i.e. how music works, how they had used music previously or how they reflected on their own actions in relation to musical habits and practices in their past or present everyday life.

The Sixth Session and Theme 'All-Time Best'

The participants chose musical materials for this theme to reinforce a conscious decision on what to choose and why. Quite often, when the participants listened to the music they had chosen, they described how they often felt sure that this was their right choice. Some participant's may choose the same piece of music they had previously chosen for previous themes. Anthony claims that 'music something that is much underrated' and how 'everything helps with music'. The theme 'All-Time

Best' contributed to an increase in self-awareness and consciousness as to how 'their' music could be a health resource and a positive contributor to well-being and 'wellness', hence part of the informal learning process:

> I think it is very exciting this consciousness around music and life, I think I have chosen music fairly unconsciously, really, so trying to relate music I choose to 'here and now' situation or even where I am… that I have not thought about as much before. [*William, age 54. Depression. Recovered prior to fieldwork. Musically active singing, playing and/or composing*]

The ability to rise above suffering, to go beyond the present situation to a realm where life takes on a deeper significance, is an important factor in the long-term management of chronic illness. Through this theme, one could detect a crucial change in Isabel's life. Her increased self-awareness of the beneficial factors of singing:

> I sing a lot, because I think it is like healing to sing. It is something that happens when you play or sing yourself, as if body and soul becomes one, one walks through a door of poetry. I always sing, when I walk, cycle or do different tasks at home. I also sing when I walk in the forest. It has been quite stormy in my life since summer, so the music has been extremely important to me, life is so strange, it does not always end up the way you intended, alas. [*Isabel, age 36, 'Burn-out syndrome'. Long-term certified sick for 16 months. Recovered during fieldwork*]

Discussion

Empowering Rituals and Informal Learning Through The Fellowship of Musicking

The interactive and dialectical use of the 'themed' music in the The Fellowship of Musicking Model may help to explore the ways that musical materials provided participants with a health technology. Firstly, this might be due to the fact that their original musical choices provided a kind of bedrock of certainty and, more specifically, aesthetic belonging, part of the process of achieving and maintaining ontological security (Giddens 1991). Through narratives of how and why some types of music seemed beneficial or not, one may observe how music can be uplifting or a highlight, could transcend or transform, often enforced through musical rituals, enhancing a strong sense of self and of fellowship, togetherness or connectedness. Ansdell (2013) describes how fellowship arises when people sing together, which may also be transferable to shared music listening. Recent empirical evidence corroborates the notion that listening to music momentarily also affects self-esteem (Elvers et al. 2015). Other research has pointed out that listening to music may elicit feelings of power, making people feel more powerful, and implicitly activate the notion of power (Hsu et al. 2014).

By playing their choices of music, which seemed to inspire them for elaborations and narratives around 'how they came to choose what they chose and why', the in-depth discussions on processes and changes in relation to music, health and quality of life may mirror their present mood and life situation. A study of everyday listening behaviour needs to take the reciprocity of the individual and the social world into account, for not only does the situation play a part in determining a person's musical preference, but the music played in a particular music situation is subsequently likely to affect behaviour (Greasley and Lamont 2006). It has been proposed elsewhere that music may serve as a resource for the regulation of mood (Saarikallio 2010) and self (Krueger 2014). While the reasons why people listen to music are manifold (Schäfer et al. 2013), the notion of musical self-enhancement addresses the specific use of music listening that is related to a gain in self-worth and the psychological processes that are involved (Elvers 2016).

One might say that one of the main features of this model to health promotion has to do with developing a ritual, a special context, in which to perform a bodily practice or a self-imaging process. This activity may lead towards control over one's life and control over the situation of being ill and being well and how one wants one's life to progress. These musical ritual practices seemed to be effective for recognizing stress, grief, anxiety, loss or fear; hence the participants seemed to constitute themselves in empowering rituals that seemed safe and secure (Fredrickson 2002). According to Maksimainen and Saarikallio (2015), the feeling of empowerment is among the three most intensely experienced emotions arising from encounters with music and art in everyday life.

From participating in The Fellowship of Musicking Model, individuals developed *skills and knowledge* on how to use music in their daily lives for enrichment and coping or as an instrument of change. The music they disapprove of and dislike or that makes them feel ill contrasts sharply with the music they 'couldn't live without'. The various stages and steps in this process highlight how participants come to produce for themselves modes of conscious awareness of music's 'powers'. Through the provision and exchange of musical works, styles and genres, this model's informal type of 'music education' or musical learning seemed to have improved individuals' health and well-being in different ways (Green 2002; Savage 2005; Batt-Rawden and DeNora 2007; DeNora 2014). Empathy has also been proposed as a potential mechanism explaining how music induces emotions in general (Clarke 2014). Teaching others and being part of a group may enable coping with stress (Lindstrøm and Eriksson 2005; Holland 1995), and 'sacred' musical moments may broaden the 'learner's' scope of cognition, enabling flexible and creative thinking (Fredrickson 2002). In line with Ruud (2005, pp. 146–7), music provides opportunities for the building of skills, for joy and pride in vocal expressions and the handling of instruments.

Participants could, by 'telling' about music's meaning and uses in their lives, create for themselves and for fellow participants pointers and tips on how to use music so as to promote health and well-being and connection to others, determining of self-conditions, self-empowerment and, furthermore, how to activate music such that it might 'work' in health-promoting or otherwise beneficial ways. For example, one could teach how music works, according to the participants, as a refresher or pick-me-up and a way of coping; how it provides relaxation and relief, vitality, personal and social competence, a sense of regaining health through playing guitar or singing; or how it soothes and calms and creates steadiness and predictability (Antonovsky 1996). Since the significance of musicking in everyday life might sometimes be buried in the depth of our subconsciousness, the fact that participants could tell how music can empower could be a door-opener into a 'new world', contributing to a deeper self-understanding and self-development, is thus a way of coping in life crises. Greasley and Lamont (2006) describe how one of their participants listens to a song on a daily basis because it helps her to find a sense of identity (Elvers 2016) and to understand herself better.

In other words, to use music in 'useful' ways involves tapping that knowledge and documenting it so as to raise it from operation to activity or as part of a reskilling process (Giddens 1991; Williams et al. 2000). The nature of lay reskilling suggests that individuals are increasingly coming to 'take back' control over matters of health and illness and returning to health and healing in an attempt to 'de-medicalize' society, in line with an increasing focus on empowerment for health. A critical reconfiguration of professional power and dominance is beginning to take place in contemporary western society (Tellnes et al. 2017).

The Fellowship of Musicking: Social Healing and Connectedness

The experiences of everyday musicking are often connected to 'high points' or peak experiences (Gabrielsson 2011; Theorell 2014) and could be sacred moments or happy memories to share among fellowmen – the 'togetherness' factor. Rituals that form musical fellowship and togetherness can be construed as a progress towards one's goals of being healed or feeling well, by creating a lifeworld that shapes ontological security (Giddens 1991) or a sense of 'being-in-time-together' (Ansdell 2004:84). These shared music-making situations seemed to promote health and well-being to a greater extent than just listening to music. An illustration of this is one participant's statement that sharing musical experience was much more healthy than 'just sitting and listening' or as another pointed out: 'you need to get physically engaged in to it, participate, you know, get involved not just listening'. The fellowship of musicking may 'lift' a person out of their everyday reality into a musical world of strong feelings or emotions inducing a sense of transcendence or affirmation (Small 1998). Personally chosen music or musicking may also create a 'key to experiencing company and coherence' (Wigram et al. 2002, p. 196).

These shared aesthetic points of a musical fellowship instigated musical narratives that pointed to the joy of being able to have somebody to share, play or sing with, for example, in a choir, a folk club or going to a musical, social event or concert, a kind of a social increasing agency. The passion for music as a bridge-builder for friendship relations and commitment is also discussed by Green (2002, p. 115). Music may provide an accessible environmental resource that allows self-worth and self-esteem to be generated by empathizing and identifying with a musical persona who shares positive attributes and a high sense of self-worth (Elvers 2016). This theme is also discussed by Bailey and Davidson (2003, 2005), Clift et al. (2008) and Balsnes (2012), who describe how singing in a choir improved the participants' self-esteem and self-confidence. As Horesh (2005) argues, music has the potential to heal and reduce stress (Theorell 2014) if one allows music to be incorporated into one's life as a source of enjoyment and enrichment, in line with a salutogenic approach to health (Antonovsky 1996).

The Significance of Context, Cohesiveness and Informal Sociality

The effects of musical participation may create a type of beneficial, social 'spin-off' effect, something to take part in each week, to learn and to look forward to, establishing a predictive ritual (Antonovsky 1996). The importance of sharing feelings and emotions through strong musical connectedness may enhance participatory consciousness (Keil and Feld 1994), which links to the value of social capital (Putnam 2000; Lin 2001; Esser 2008). In line with Green (2002), these informal learning situations may then highlight the importance of participation when it comes to the core of the beneficial factors of deep (self-) musicking, the togetherness factor, the embodied and special kind of 'feelingful' activity (Keil and Feld 1994). That said, previous research (Batt-Rawden 2007) shows how participants organized themselves into a 'music' group after the project, expressing the importance of belonging to a musical group, hence the importance for public health to create 'stepping stones' for future, informal music-making in the local communities. The social connectedness in turn may enhance our relational value and interpersonal acceptance and thus enhance our self-esteem and self-worth (Elvers 2016). Ansdell (2004:83–86) adds the concept of *musical communitas* to a common shared world of time, space, gesture and energy which allows both diversity and unity; that is, the particular possibilities and qualities of social and cultural experience are motivated and sustained through musicking. Thus, shared and focused musicking will strengthen cohesiveness and togetherness in a group (Theorell 2014).

Final Comments

Informal 'Health-Promoting' Teaching Cultures

The Fellowship of Health Musicking Model is one way to create informal health-promoting cultures of learning or teaching cultures', increasing practical (active) musicking involvement for the individuals in local communities, thus building social identities (Tarrant et al. 2001; Lonsdale and North 2009) and social capital

(Putnam 2000; Esser 2008). The fact that they have been part of a group gives them an opportunity to form their own group or subgroup(s) if they wish. Close social networks have been well documented to have a huge impact on our health and quality of life and to be an important part of our mental and social well-being (Putnam 2000; Sørensen et al. 2011; DeNora 2000, Ruud 2001; Batt-Rawden and Tellnes 2005).

The Fellowship of Musicking Model is suitable to be adopted or further developed for promoting salutogenic musical activities involving members in local communities (Koelen et al. 2005; Small 1998; Stige 2005, 2012; Bonde 2011; Green 2002) and a way to increase a 'participatory consciousness' as ecological beings (Keil and Feld 1994, p. 97). Another key function of music-making and listening that has been discussed in relation to its use today as well as its evolutionary development concerns its capacity to create and strengthen social bonds among those who interact and engage in musical activities (Boer 2011; Tarr et al. 2014). Local communities may be seen as an arena for identity building – a site for co-learning and empowering processes (Koelen et al. 2005; MacDonald et al. 2002).

Recommendation on How to Implement the Model in Different Contexts

The Coordination Reform (2009)[9] together with a new Public Health Law (2012)[10] from Norway emphasizes challenges related to patients' needs for coordinated services, as well as greater efforts to prevent disease and promote health. As such, new approaches and methods in health promotion are needed to meet future public health issues and challenges (Tellnes et al. 2017). Cross actors' [here facilitators] cooperation can result in new and innovative solutions to welfare and health problems, and a recent study relating to public and private innovation shows the importance of municipalities' acceptance of new ideas as good or useful (Fuglesang et al. 2015).

Public health research and practice should focus not only on factors causing disease and injuries (pathogenesis) but also factors promoting health (salutogenesis) in the perspective of health promotion and prevention in different contexts.

To sum up, there are several contexts and settings where this model could be implemented by different facilitators. GPs in cooperation with social workers, nurses and/or public health officers in local communities could initiate the model as an alternative approach to promote health for men and women with long-term illnesses and therefore motivate participants to return to work or to cope with health challenges and/or to change lifestyle towards healthy behaviours (Batt-Rawden and

---

[9] White Paper. [St.meld. nr. 47 (2008–2009)]. Coordination Reform, Norway. [Samhandlingsreformen— Rett behandling – på rett sted – til rett tid] https://helsedirektoratet.no/samhandlingsreformen.

[10] Public Health Law (2012). https://lovdata.no/dokument/NL/lov/2011-06-24-29.

Tellnes 2012). As noted by Bonde (2011, p. 134), a lifestyle-based paradigm music intervention can be part of lifestyle enhancement programmes that not only teach strategies for coping with stress and pain but also support strategies for breaking insufficient cognitive, emotional and relational patterns.

Thus, the model could also be used by facilitators in hospitals, institutions, rehabilitation centres and/or local communities, namely, as a follow-up procedure for individuals after a period of sickness absence or long-term illness. A recent study from Norway (Batt-Rawden et al. 2017) shows how private rehabilitation centres often face a common challenge regarding the high risk of relapse after treatment and rehabilitation among patient groups with disabilities, psychiatric disorders, mental disorders, obesity and drug addiction. There is a need to cooperate with public health officers, nurses, social workers, GPs and other professionals with similar qualifications.

Furthermore, according to new regulations (Directory of Health Norway 2013), this model could also be adopted or further developed by music therapists working in psychiatric institutions, since there is a call for using music as an alternative or complimentary approach to treatment and less medication. Moreover, nurses could be taught how to incorporate music interventions for individualized patient care into their practice to more effectively manage anxiety and regulate pain, bodily tension and stress. This model could also be well worth exploring or trying out in relation to specific groups of illnesses, diseases or disorders (e.g. anxiety/depression, muscular disease, cancer or cardio-vascular disease), subcultures or age groups. However, this is still to be explored and documented.

It must be stressed that health musicking is not limited to a professional therapeutic context but can be observed in any social or individual practice where people use music experiences to create meaning and coherence in states and times of adversity. To emphasize this point, the interdisciplinary field of health musicking includes professionals with many different backgrounds and qualifications (Bonde 2011, p. 135). To highlight The Fellowship of Musicking Model, it is a salutogenic activity for health maintenance and transformation related to public health issues.

References

Ansdell, G. (2004). Rethinking music and community: Theoretical perspectives in support of community music therapy. In M. Pavlicevic & G. Ansdell (Eds.), *Community music therapy*. London: Jessica Kingsley Publishers.

Ansdell, G. (2013). Foreword: to music's health. In E. Bonde, E. Ruud, M. S. Skånland, & G. Trondalen (Eds.), *Musical life stories: Narratives on health musicking* (Vol. 5). Oslo: NMH-Publikasjoner.

Ansdell, G. (2015). *How music helps in music therapy and everyday life*. Surrey: Ashgate.

Antonovsky, A. (1996). The salutogenic model as a theory to guide health promotion. *Health Promotion International, 11*(1), 11–18.

Bailey, B. A., & Davidson, J. W. (2003). Amateur group singing as a therapeutic instrument. *Nordic Journal of Music Therapy, 12*(1), 18–32.

Bailey, B., & Davidson, J. (2005). Effects of group singing and performance for marginalized and middle-class singers. *Psychology of Music, 33*(3), 269–303.
Balsnes, A. H. (2012). Choral singing, health and quality of life: The story of Diana. *Arts and Health, 4*(3), 249–261.
Batt-Rawden, K. B. (2007). *Music and Health Promotion: The role and significance of music and musicking in the lives of men and women with long term illnesses* (Doctoral thesis). University of Exeter, England.
Batt-Rawden, K. B. (2010). The benefits of self-selected music on health and well being. *The Arts in Psychotherapy, 37*, 301–310.
Batt-Rawden, K. B. Bjørk E., & Waaler, D. (2017). Human factors in implementation and adoption ofinnovations in health care services: A longitudinal case study on the introduction of new technology. *The Innovation Journal: The Public Sector. 22*(3), 2017, article 3.
Batt-Rawden, K.B., & Aasgaard, T. (2006). Music a key to the kingdom. *Electronic Journal of Sociology*. http://www.sociology.org/content/2006/tier1/batt-rawden.html. ISSN - 1198 3655: pp. 1–21. Retrieved online 8th of June, 2006.
Batt-Rawden, K. B., & DeNora, T. (2007). Music and informal learning in everyday life. *Music Education Research, 7*(3), 289–304.
Batt-Rawden, K. B., & Tellnes, G. (2005). Nature-culture-health activities as a method of rehabilitation; an evaluation of participants' health, quality of life and function. *International Journal of Rehabilitation Research, 28*, 175–180.
Batt-Rawden, K. B., & Tellnes, G. (2011). How music may promote healthy behaviors. *Scandinavian Journal of Public Health, 39*, 113–120.
Batt-Rawden, K. B., & Tellnes, G. (2012). Social factors of sickness absences and ways of coping: A qualitative study of men and women with mental and musculoskeletal diagnoses. *International Journal of Mental Health Promotion, 14*, 83–95.
Batt-Rawden, K. B., DeNora, T., & Ruud, E. (2005). Music listening and empowerment in health promotion: A study of the role and significance of music in everyday life of the long-term ill. *Nordic Journal of Music Therapy, 14*(2), 120–136.
Batt-Rawden, K. B., Trythall, S., & DeNora, T. (2007). Health musicking as cultural inclusion. In J. Edwards (Ed.), *Music: Promoting health and creating community in healthcare contexts* (pp. 64–82). Cambridge, UK: Cambridge Scholars Publishing.
Baumgartner, H. (1992). Remembrance of things past: Music, autobiographical memory, and emotion. *Advances in Consumer Research, 19*, 613–620.
Becker, J. (2004). *Deep listeners: Music, emotion and trancing*. Bloomington: Indiana University Press.
Boer, D. (2011). How share preferences in music create bonds between people: Values as the missing link. *Personality and Social Psychology Bulletin, 37*, 1159–1171. https://doi.org/10.1177/0146167211407521.
Bonde, L. O. (2011). Health music(k)ing – music therapy or music and health? A model, eight empirical examples and some personal reflections". *Music and Arts in Action. Special Issue: Health Promotion and Wellness, 140*, 126–140.
Bonde, L. O., Ruud, E., Skånland, M. S., & Trondalen, G. (Eds.). (2013). *Musical life stories. Narratives on health musicking* (Vol. 5). Oslo: NMH-publikasjoner.
Brunges, M., & Avigne, R. N. (2003). Music therapy for reducing surgical anxiety. *AORNJ, 78*(5), 816–818. http://gateway.ut.ovid/com/gw1/ovidweb.cgi. [retrieved online 7th of January 2004].
Butterton, M. (2004). *Music and meaning. Opening minds in the caring and healing professions*. Oxon: Radcliff Medical Press.
Chanda, M. L., & Levitin, D. J. (2013). The neurochemistry of music. In: *Trends in Cognitive Science, 17*(4), 179–194.
Charmaz, K. (1999). Stories of suffering: Subjective tales and research narratives. *Qualitative Health Research, 9*(3), 362–382.
Clarke, E. (2014). Lost and found in music: Music, consciousness and subjectivity. *Musicae Scientiae, 18*, 354–368. https://doi.org/10.1177/1029864914533812.

Clift, S., Hancox, G., Staricoff, R., & Whitmore, C. (2008). A systematic mapping and review of non-clinical research on singing and health. *Sidney De Haan Research Centre for Arts and Health*, Canterbury: Canterbury Christ Church University, UK.

Cuypers, K., Krokstad, S., & Holmen, T. L. (2013). Patterns of receptive and creative cultural activities and their association with perceived health, anxiety, depression and satisfaction with life among adults: The HUNT study, Norway. *Journal of Epidemiological Community Health, 66*, 698–703.

DeNora, T. (1999). Music as a technology of the self. *Poetics, 27*(1), 31–56.

DeNora, T. (2000). *Music in everyday life*. Cambridge: Cambridge University Press.

DeNora, T. (2014). *Music asylums: Wellbeing through music in everyday life*. London: Routledge.

Directory of Health [Helsedirektoratet]. (2013). National academic guidelines for the investigation, treatment and follow-up of persons with psychosis disorders. [Nasjonal faglig retningslinje for utredning, behandling og oppfølging av personer med psykoselidelser]. *Directory of Health. IS-number: IS-1957*. ISBN-nr. 978-82-8081-242-1.

Elvers, P. (2016). Songs for the ego: Theorizing musical self-enhancement. *Frontiers in Psychology, 7*, 2. https://doi.org/10.3389/fpsyg.2016.00002.

Elvers, P., Fischinger, T., & Steffens, J. (2015). Listening to empowering music: Music induced manipulations of self-esteem. *Paper Presented at the Ninth Triennial Conference of the European Society for the Cognitive Sciences of Music*, Manchester.

Esser, H. (2008). The two meanings of social capital. In D. Castiglione, J. W, V. Deth, & W. Guglielmo (Eds.), *The handbook of social capital* (pp. 22–49). Oxford: Oxford University Press.

Fancourt, D., Ockelford, A., & Belai, A. (2014). The psychoneuroimmunological effects of music: A systematic review and a new model. *Brain, Behavior, Immunity, 6*, 15–26.

Fredrickson, B. L. (2002). Positive emotions. In C. R. Snyder & S. Lopez (Eds.), *Handbook of positive psychology*. Oxford: Oxford University Press.

Fuglesang, L., Hulgård, L., & Langergaard, L. (2015). An updated list of Municipal public-private innovation studies in Europe, especially Nordic studies. *Prepared for the research project Municipal Innovation Research for Institutional Development (MIRID), WP1*, at Lillehammer University College. Roskilde University.

Gabrielsson, A. (2011). *Strong experiences with music*. Oxford: Oxford University Press.

Gagner-Tjellesen, D., Yurkovich, E. E., & Gragert, M. (2001). The use of music therapy and other ITNIs in acute care. *Journal of Psychosocial Nursing, 39*(10), 26–37. 52–3.

Giddens, A. (1991). *Modernity and self-identity*. Cambridge: Polity Press.

Greasley, A. E., & Lamont, A.M. (2006). Music preferences in adulthood: Why do we like the music we do?. In M. Baroni, et al. (Ed.), Proceedings of the 9th International Conference on Music Perception & Cognition (ICNPC9). *The Society for Music Perception and Cognition (SMPC) and European Society for the Cognitive Sciences of music (ESCOM)*. Bologna/Italy, Aug 22–26 2006, 960.

Green, L. (2002). *How popular musicians learn. A way ahead for music education*. Aldershot: Ashgate.

Hansen, E., Sund, E., & Krokstad, S. (2015). Cultural activity participation and associations with self-perceived health, life-satisfaction and mental health: The young HUNT study Norway. *BMC Public Health, 15*, 544.

Holland, P. (1995). The role of music in the effective relief of stress. In T. Wigram, B. Saperston, & R. West (Eds.), *The art and science of music therapy: A handbook* (pp. 406–432). Longshore: Harwood Academic Press.

Horesh, T. (2005). Dangerous music – working with the destructive and healing powers of popular music in the treatment of substance abusers. In: D. Aldridge & J. Fachner (Eds.), *Music and Altered States: Consciousness, Transcendence, Therapy and Addiction*. pp: 125–140. London: Jessica Kingsley Publishers.

Hsu, D. Y., Huang, L., Nordgren, L. F., Rucker, D. D., & Galinsky, A. D. (2014). The music of power: Perceptual and behavioral consequences of powerful music. *Social Psychological and Personality Science, 6*, 75–83. https://doi.org/10.1177/1948550614542345.

Juslin, P. N., & Sloboda, J. A. (Eds.). (2010). *Handbook of music and emotion*. NewYork: Oxford University Press.
Keil, C., & Feld, S. (1994). *Music grooves*. Chicago: The University of Chicago Press.
Koelen, M., Vaandrager, L., & Colomér, C. (2005). Health promotion research: Dilemmas and challenges. *Journal of Epidemiological Community Health, 55*(4), 257–262.
Konlaan, B. B., Bygren, L. O., & Johansson, S. E. (2000). Visiting cinema, concerts, museums or art exhibitions as dominant of survival: A Swedish fourteen-year cohort follow up. *Scandinavian Journal of Public Health, 28*, 174–178.
Krueger, J. (2014). Affordances and the musically extended mind. *Frontiers in Psychology, 4*, 1003. https://doi.org/10.3389/fpsyg.2013.01003.
Langeland, E., & Wahl, A. K. (2009). The impact of social support on mental health service users' sense of coherence: A longitudinal panel survey. *International Journal of Nursing Studies, 46*, 830–837.
Langeland, E., Wahl, A. K., Kristoffersen, K., & Hanestad, B. R. (2007). Promoting coping: Salutogenesis among people with mental health problems. *Issues in Mental Health Nursing, 28*, 275–295.
Lin, N. (2001). *Social capital. A theory of social structure and action*. Cambridge: Cambridge University Press..
Lindstrøm, B., & Eriksson, M. (2005). Salutogenesis. *Journal of Epidemiological Community Health, 59*(6), 440–442.
Lindstrøm, B., & Eriksson, E. (2015). *Haikerens guide til salutogenese. [Hitchhaikers Guide to Salutogenesis]*. Oslo: Gyldendal Akademisk.
Lonsdale, A. J., & North, A. C. (2009). Musical taste and ingroup favouritism. Group process. *Intergroup Relations, 12*, 319–327. https://doi.org/10.1177/1368430209102842.
MacDonald, D. J., Hargreaves, D., & Miell, D. (Eds.). (2002). *Musical identities*. Oxford: Oxford University Press.
Maksimainen, J., & Saarikallio, S. (2015). Affect from art: subjective constituents of everyday pleasure of music and pictures: Overview and early results. *Paper Presented at the Ninth Triennial Conference of the European Society for the Cognitive Sciences of Music*, Manchester.
Maranto, C. D. (1993). Applications of music in medicine. In M. Heal & T. Wigram (Eds.), *Music therapy in health and education*. London: Jessica Kingsley Publishers.
Mittelmark, M. B., Sagy, S., Eriksson, M., Bauer, G., Pelikan, J. M., Lindström, B., et al. (2016). *The handbook of Salutogenesis*. Springer.
Napier, D., et al. (2011). Culture and health lancet commission. *The Lancet, 384*, 1607–1639.
Putnam, R. D. (2000). *Bowling alone. The collapse and revival of American community*. New York: Simon and Schuster.
Ruud, E. (2001). *Varme øyeblikk. [Happy moments]*. Oslo: Unipub.
Ruud, E. (2002). Music as a cultural immunogen – three narratives on the use of music as a technology of health. In I. M. Hanken, S. G. Nilsen, & M. Nerland (Eds.), *Research in and for higher music education. Festschrift for Harald Jørgensen* (Vol. 2). Oslo: NMH-Publications.
Ruud, E. (2005). Music: A salutogenic way to health promotion? In G. Tellnes (Ed.), *Urbanization and health. New challenges to health promotion and prevention*. Oslo: Academic, UniPub.
Saarikallio, S. (2010). Music as emotional self-regulation throughout adulthood. *Psychology of Music, 39*, 307–327. https://doi.org/10.1177/0305735610374894.
Savage, J. (2005). Sound2Picture: Developing compositional pedagogies from the sound designer's world. *Music Education Research, 7*(3), 331–348.
Schäfer, T., Sedlmeier, P., Stadtler, C., & Huron, D. (2013). The psychological functions of music listening. *Frontiers in Psychology, 4*, 511. https://doi.org/10.3389/fpsyg.2013.00511.
Sloboda, J. A. (2005). *Exploring the musical mind*. Oxford: Oxford University Press.
Small, C. (1998). *Musicking. The meanings of performing and listening*. London: Weslyan.
Sørensen, T., Klungsoyr, O., Kleiner, R., & Klepp, O. M. (2011). Social support and sense of coherence: Independent, shared and interaction relationships with life stress and mental health. *The International Journal of Mental Health Promotion, 13*, 27–44.

Stige, B. (2005). Music as a health resource. *Nordic Journal of Music Therapy, 14*(1). http://www.hisf.no/njmt/editorial141.html. Retrieved online 2nd of Nov 2005.

Stige, B. (2012). Health musicking: A perspective on music and health as action and performance. In R. MacDonald, G. Kreutz, & L. Mitchell (Eds.), *Music, health and wellbeing* (pp. 183–195). Oxford: Oxford University Press.

Tarr, B., Launay, J., & Dunbar, R. I. (2014). Music and social bonding: "Self– other" merging and neurohormonal mechanisms. *Frontiers in Psychology, 5*, 1096. https://doi.org/10.3389/fpsyg.2014.01096.

Tarrant, M., North, A. C., & Hargreaves, D. J. (2001). Social categorization, self-esteem, and the estimated musical preferences of male adolescent. *Journal of Social Psychology, 141*, 565–581. https://doi.org/10.1080/00224540109600572.

Tellnes, G. (2003). Public health and the way forward. In W. Kirch (Ed.), *Public health in Europe*. Berlin: Springer-Verlag.

Tellnes, G. (2005). Health promotion in the local communities. (President's column). *European Journal of Public Health, 15*, 331.

Tellnes, G. (Ed.) (2017). *Nature-culture-health in public health; health promoting collaboration, prevention and rehabilitation [Natur og kultur som folkehelse: Helsefremmende samhandling, forebygging og rehabilitering]*. Oslo: Fagbokforlaget.

The Norwegian Public Health Act. Ministry of Health and Care Services (MHCS, Norway). http://app.uio.no/ub/ujur/oversatte-lover/data/lov-20110624-029-eng.pdf.

Theorell, T. (2014). *Psychological health effects of musical experiences: Theories, studies and reflections in music health science*. New York: Springer.

Wigram, T., Nygaard Pedersen, I., & Bonde, L. O. (2002). *A comprehensive guide to music therapy*. London: Jessica Kingsley Publishers.

Williams, S. J., Gabe, J., & Calnan, M. (Eds.). (2000). *Health, medicine and society: Key theories, future agendas*. London: Routledge.

Index

L. O. Bonde, T. Theorell (eds.), *Music and Public Health*,
https://doi.org/10.1007/978-3-319-76240-1

L

M

The manufacturer's authorised representative in the EU is Springer Nature Customer Service Centre GmbH, Europaplatz 3, 69115 Heidelberg, Germany. If you have any concerns regarding our products, please contact ProductSafety@springernature.com

Printed and bound by CPI Group (UK) Ltd, Croydon, CR0 4YY
06/10/2026
02228973-0002